Electric Transmission of Water Power

Alton D. Adams

Classical Prints

This edition published in 2022

ISBN : 9789356535961

Contents

CHAPTER I.
WATER-POWER IN ELECTRICAL SUPPLY.

Electrical supply from transmitted water-power is now distributed in more than fifty cities of North America. These include Mexico City, with a population of 402,000; Buffalo and San Francisco, with 352,387 and 342,782 respectively; Montreal, with 266,826, and Los Angeles, St. Paul, and Minneapolis, with populations that range between 100,000 and 200,000 each. North and south these cities extend from Quebec to Anderson, and from Seattle to Mexico City. East and west the chain of cities includes Portland, Springfield, Albany, Buffalo, Hamilton, Toronto, St. Paul, Butte, Salt Lake City, and San Francisco. To reach these cities the water-power is electrically transmitted, in many cases dozens, in a number of cases scores, and in one case more than two hundred miles. In the East, Canada is the site of the longest transmission, that from Shawinigan Falls to Montreal, a distance of eighty-five miles.

From Spier Falls to Albany the electric line is forty miles in length. Hamilton is thirty-seven miles from that point on the Niagara escarpment, where its electric power is developed. Between St. Paul and its electric water-power station, on Apple River, the transmission line is twenty-five miles long. The falls of the Missouri River at Cañon Ferry are the source of the electrical energy distributed in Butte, sixty-five miles away. Los Angeles draws electrical energy from a plant eighty-three miles distant on the Santa Ana River. From Colgate power-house, on the Yuba, to San Francisco, by way of Mission San José, the transmission line has a length of 220 miles. Between Electra generating station in the Sierra Nevada Mountains and San Francisco is 154 miles by the electric line.

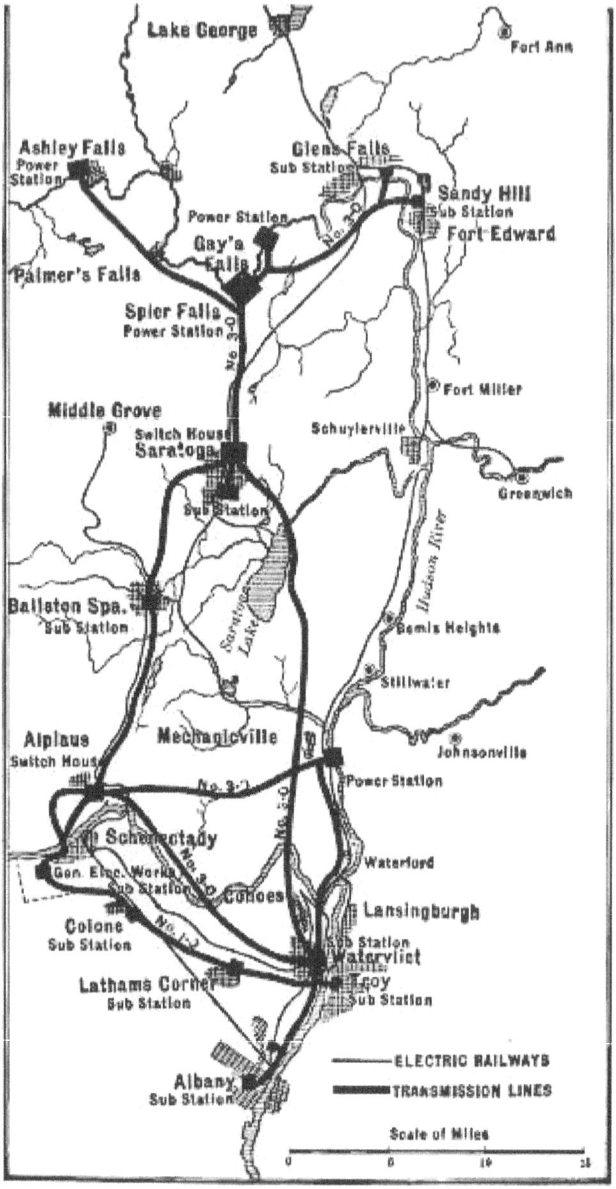

FIG. 1.—Spier Falls Transmission Lines.

These transmissions involve large powers as well as long distances. The new plant on the Androscoggin is designed to deliver 10,000 horse-power for electrical supply in Lewiston, Me. At Spier Falls, on the Hudson, whence energy goes to Albany and other cities, the electric generators will have a capacity of 32,000 horse-power. From the two water-power stations at Niagara Falls, with their twenty-one electric generators of 5,000 horse-power each, a total of 105,000, more than 30,000 horse-power is regularly transmitted to Buffalo alone; the greater part of the capacity being devoted to local industries. Electrical supply in St. Paul is drawn from a water-power plant of 4,000 and in Minneapolis from a like plant of 7,400 horse-power capacity. The Cañon Ferry station, on the Missouri, that supplies electrical energy in both Helena and Butte, has a capacity of 10,000 horse-power. Both Seattle and Tacoma draw electrical supply from the 8,000 horse-power plant at Snoqualmie Falls. The Colgate power-house, which develops energy for San Francisco and a number of smaller places, has electric generators of 15,000 horse-power aggregate capacity. At the Electra generating station, where energy is also transmitted to San Francisco and other cities on the way, the capacity is 13,330 horse-power. Electrical supply in Los Angeles is drawn from the generating station of 4,000 horse-power, on the Santa Ana River, and from two stations, on Mill Creek, with an aggregate of 4,600, making a total capacity of not less than 8,600 horse-power. Five water-power stations, scattered within a radius of ten miles and with 4,200 horse-power total capacity, are the source of electrical supply in Mexico City.

The foregoing are simply a part of the more striking illustrations of that development by which falling water is generating hundreds of thousands of horse-power for electrical supply to millions of population. This application of great water powers to the industrial wants of distant cities is hardly more than a decade old. Ten years ago Shawinigan Falls was an almost unheard-of point in the wilds of Canada. Spier Falls was merely a place of scenic interest; the Missouri at Cañon Ferry was not lighting a lamp or displacing a pound of coal; that falling water in the Sierra Nevada Mountains should light the streets and operate electric cars in San Francisco seemed impossible, and that diversion of Niagara, which seems destined to develop more than a million horse-power and leave dry the precipices over which the waters now plunge, had not yet begun. In some few instances where water-power was located in towns or cities, it has been applied to electrical supply since the early days of the industry. In the main, however, the supply of electrical energy from water-power has been made possible only by long-distance transmission. The extending radius of electrical transmission for water-powers has formed the greatest incentive to their development. This development in turn has reacted on the conditions that limit electrical supply and has materially extended the field of its application. Transmitted water-power has reduced the rates for electric service. It may not be easy to prove this reduction by quoting figures for net rates, because these are not generally published, but there are other means of reaching the conclusion.

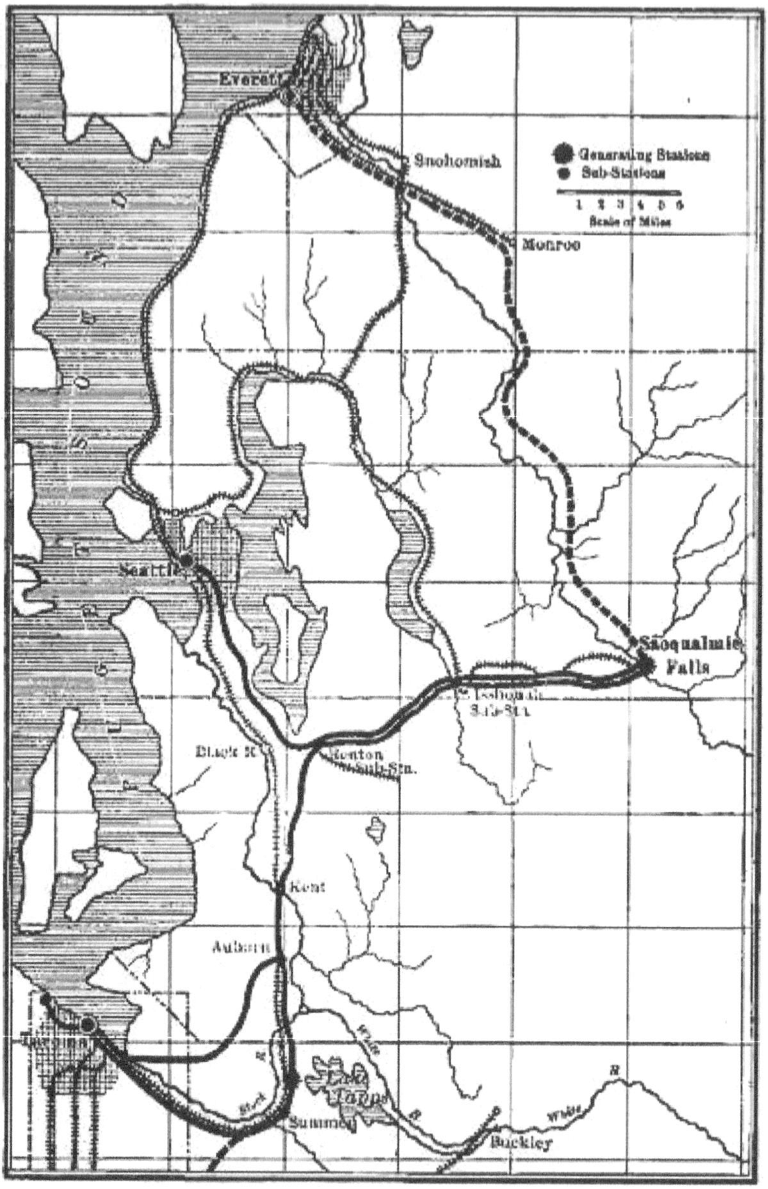

FIG. 2.—Snoqualmie Falls Transmission Lines.

In the field of illumination electricity competes directly with gas, and in the field of motive power with coal. During the past decade it is well known that the price of gas has materially declined and the price of coal, barring the recent strike period, has certainly not increased. In spite of these reductions electrical supply from water-power has displaced both gas and coal in many instances.

Moreover, the expansion of electric water-power systems has been decidedly greater, as a rule, than that of electrical supply from steam-driven stations. An example of the fact last stated may be seen in Portland, Me. In the spring of 1899, a company was formed to transmit and distribute electrical energy in that city from a water-power about thirteen miles distant. For some years, prior to and since the date just named, an extensive electric system with steam-power equipment has existed in Portland. In spite of this, the system using water-power, on January 1st, 1903, had a connected load of 352 enclosed arcs and 20,000 incandescent lamps, besides 835 horse-power in motors.

Comparing the expansion of electric water-power systems with those operated by steam, when located in different cities, Hartford and Springfield may be taken on the one hand and Fall River and New Bedford on the other. The use of water-power in electrical supply at Hartford began in November, 1891, and has since continued to an increasing extent. Throughout the same period electrical supply in Fall River has been derived exclusively from steam. In 1890 the population of Hartford was 53,230, and in 1900 it stood at 79,850, an increase of 50 per cent. At the beginning of the decade Fall River had a population of 74,398, and at its close the figures were 104,863, a rise of 40.9 per cent. In 1892 the connected load of the electric supply system at Fall River included 451 arc and 7,800 incandescent lamps, and motors aggregating 140 horse-power. By 1901 this load had increased to 1,111 arcs, 24,254 incandescent lamps, and 600 horse-power in motors. The electric supply system at Hartford in 1892 was serving 800 arcs, 2,000 incandescent lamps, and no motors. After the use of transmitted water-power during nine years the connected load of the Hartford system had come to include 1,679 arcs, 68,725 incandescent lamps, and 3,476 horse-power of motor capacity in 1901. At the beginning of the decade Hartford was far behind Fall River in both incandescent lamps and motors, but at the end Hartford had nearly three times as many incandescent lamps and nearly six times as great a capacity in connected motors. As Fall River had a population in 1900 that was greater by thirty-one per cent. than the population of Hartford, and the percentage of increase during the decade was only 9.1 lower in the former city, water-power seems to have been the most potent factor in the rise of electric loads in the latter. Electric gains at Hartford could not have been due to the absence of competition by gas, for the price of gas there in 1901 was $1 per 1,000 cubic feet, while the price in Fall River was $1.10 for an equal amount.

Water-power began to be used in electrical supply at Springfield during the latter half of 1897. In that year the connected load of the Springfield electric system included 1,006 arcs, 24,778 incandescent lamps, and motors with a capacity of 647 horse-power. Five years later, in 1902, this connected load had risen to 1,399 arc lamps, 45,735 incandescent lamps, and a capacity of 1,025 horse-power in electric motors. At New Bedford, in 1897, the electric system was supplying 406 arc and 22,122 incandescent lamps besides motors rated at 298 horse-power. This load, in 1902, had changed to 488 arcs, 18,055 incandescent lamps, and 432 horse-power in capacity of electric motors. From the foregoing figures it appears that while 82 arc lamps were added in New Bedford, 393 such lamps were added in Springfield. While the electric load at New Bedford was increased by 134 horse-power of motors, the like increase at Springfield was 378 horse-power, and while the former city lost 4,067 from its load of incandescent lamps, the latter gained 20,957 of these lamps. During all these changes electrical supply in Springfield has come mostly from water-power, and that in New Bedford has been the product of steam. Population at Springfield numbered 44,179 in 1890 and 62,059 in 1900, an increase of 40.5 per cent. In the earlier of these years New Bedford had a population of 40,733, and in the later 62,442, an increase of 53.3 per cent. In 1902 the average price obtained for gas at Springfield was $1.04 and at New Bedford $1.18 per 1,000 cubic feet.

Springfield contains a prosperous gas system, and the gross income there from the sale of gas was thirty-one per cent greater in 1902 than in 1897. During this same period of five years the gross income from sales of electrical energy, developed in large part by water-power, increased forty-seven per cent. For the five years of general depression, ending in 1897 gross annual income of gas sales in Springfield rose only five per cent, and the like electric income nine per cent. In the five years last named the electrical supply system was operated with coal.

The application of transmitted water-power in electrical supply has displaced steam as a motive power in many large industrial plants that never would have been operated from steam-driven electric stations. An example of this sort exists at Portland, where one of the motors operated by the electric water-power system, in an industrial plant, has a capacity of 300 horse-power. Every pound of coal burned in Concord, N. H., is hauled by the single steam railway system entering that city, which railway operates large car and repair shops there. Some years ago the railway installed a complete plant of engines, dynamos, and motors for electric-driving throughout these shops. These engines and dynamos now stand idle and the motor equipment, with an aggregate capacity of 590 horse-power, is operated with energy purchased from the local electrical supply system and drawn from water-power.

Another striking example of the ability of electric water-power systems to make power rates that are attractive to large manufacturers may be seen at

Manchester, N. H. One of the largest manufacturing plants in that city purchases energy for the operation of the equivalent of more than 7,000 incandescent lamps, and of motors rated at 976 horse-power, from the electrical supply system there, whose generating stations are driven mainly by water-power. The Manchester electrical supply system also furnishes energy, through a sub-station of 800-horse-power capacity, to operate an electric railway connecting Manchester and Concord. This electric line is owned and operated in common with the only steam railway system of New Hampshire, so that the only inducement to purchase energy from the water-power system seems to be one of price.

In Buffalo the electric transmission system from Niagara Falls supplies large motors of about 20,000 horse-power capacity in manufacturing and industrial works, and 7,000 horse-power to the street railway system, besides another 4,000 horse-power for general service in lighting and small motors. Few large cities in the United States have cheaper coal than Buffalo, and in Portland, Concord, and Manchester coal prices are moderate. In the Rocky Mountain region, where coal is more expensive, the greater part of the loads of some electric water-power systems is made up of large industrial works. In Salt Lake City the electrical supply system, which draws its energy almost exclusively from water-powers, had a connected load of motors aggregating 2,600 horse-power as far back as 1901, and also furnished energy to operate the local electric railway, and several smelters six miles south of the city, besides all the local lighting service. As good lump coal sells in Salt Lake for $4.50 per ton, slack at less than one-half this figure, and the population there by the late census was only 53,531, the figures for the load of motors are especially notable. At Helena energy from the 10,000 horse-power station at Cañon Ferry operates the local lighting and power systems, two smelting and a mining plant.

Cities with Electrical Supply from Water-Power.

City.	Miles from Water-Power to City.	Horse-Power of Water-Driven Stations.	Population.
Mexico City	10 to 15	4,200	402,000
Buffalo	23	[A]30,000	352,387
Montreal	85	—	266,826
San Francisco	147	13,330	342,782
Minneapolis	10	7,400	202,718
St. Paul	25	4,000	163,065
Los Angeles	83	8,600	102,479
Albany	40	32,000	94,151
Portland, Ore.	—	—	90,426
Hartford	11	3,600	79,850
Springfield, Mass.	6	3,780	62,059
Manchester, N. H.	13 .5	5,370	59,987
Salt Lake City	36 .5	10,000	53,531
Portland, Me.	13	2,660	50,145
Seattle	—	8,000	80,671
Butte	65	10,000	30,470
Oakland	142	15,000	66,900
Lewiston, Me.	3	3,000	23,761
Concord, N. H.	4	1,000	19,632
Helena, Mont.	20	—	10,770
Hamilton, Ont.	35	8,000	
Quebec	7	3,000	
Dales, Ore.	27	1,330	

[A] Power received.

In Butte, energy from the station just named operates the works of five smelting and mining companies, driving motors that range from 1 to 800 horse-power in individual capacity. The capacity of the Butte sub-station is 7,600 horse-power.

The great electric water power system marked by the Santa Ana station at one end and the city of Los Angeles at the other, eighty-three miles distant, includes more than 160 miles of transmission lines, several hundred miles of distribution circuits, and supplies light and power in twelve cities and towns. Among the customers of this system are an electric railway, a number of irrigation plants, and a cement works. These works contain motors that range from 10 to 200 horse-power each in capacity. Motors of fifty horse-power or less are used at pumping stations in the irrigation systems.

Applications of water-power in electrical supply during the past decade have prepared the way for a much greater movement in this direction. Work is now under way for the electric transmission of water-power, either for the first time or in larger amounts, to Albany, Toronto, Chicago, Duluth, Portland, Oregon, San Francisco, Los Angeles, and dozens of other cities that might be named.

Another ten years will see the greater part of electrical supply on the American continent drawn from water-power.

Only the largest city supplied from each water-power is named above. Thus the same transmission system enters Albany, Troy, Schenectady, Saratoga, and a number of smaller places.

CHAPTER II.

UTILITY OF WATER-POWER IN ELECTRICAL SUPPLY.

In comparatively few systems is the available water-power sufficient to carry the entire load at all hours of the day, and during all months of the year, so that the question of how much fuel can be saved is an uncertain one for many plants. Again, the development of water-power often involves a large investment, and may bring a burden of fixed charges greater than the value of the fuel saved.

In spite of these conflicting opinions and factors, the application of water-power in electrical systems is now going on faster than ever before. If a saving of fuel, measured by the available flow of water during those hours when it can be devoted directly to electrical supply, were its only advantage, the number of cases in which this power could be utilized at a profit would be relatively small. If, on the other hand, all of the water that passes down a stream could be made to do electrical work, and if the utilization of this water had other advantages nearly or quite as great as the reduction of expense for coal, then many water-powers would await only development to bring profit to their owners.

No part of the problem is more uncertain than the first cost and subsequent fixed charges connected with the development of water-power. To bring out the real conditions, the detailed facts as to one or more plants may be of greater value than mere general statements covering a wide range of cases.

On a certain small river the entire water privilege at a point where a fall of fourteen feet could be made available was obtained several years ago. At this point a substantial stone and concrete dam was built, and also a stone and brick power-house with concrete floor and steel truss roof. In this power-house were installed electric generators of 800 kilowatts total capacity, direct-connected to horizontal turbine wheels. The entire cost of the real estate necessary to secure the water-power privilege plus the cost of all the improvements was about $130,000. More than enough water-power to drive the 800-kilowatt generators at full load was estimated to be available, except at times of exceptionally low water. At this plant the investment for the water-power site, development, and complete equipment was thus $162 per kilowatt capacity of generators installed.

Allowing 65 days of low water, these generators of 800 kilowatts capacity may be operated 300 days per year. If the running time averages ten hours daily at full load, the energy delivered per year is 2,400,000 kilowatt hours. Ten per cent of the total investment should be ample to cover interest and depreciation charges, and this amounts to $13,000 yearly. It follows that the items of interest and depreciation on the original investment represent a charge of 0.54 cent per kilowatt hour on the assumed energy output at this plant. This energy is transmitted a few miles and used in the electrical supply system of a large city.

On another river the entire water privilege was secured about four years ago at a point where a fall of more than 20 feet between ledges of rock could be obtained and more than 2,000 horse-power could be developed. At this point a masonry dam and brick power-house were built, and horizontal turbine wheels were installed, direct-connected to electric generators of 1,500 kilowatts total capacity. The entire cost of real estate, water rights, dam, building, and equipment in this case was about $250,000.

Assuming, as before, that generators may be operated at full capacity for 10 hours per day during 300 days per year, the energy delivered by this plant amounts to 4,500,000 kilowatt hours yearly. The allowance of 10 per cent on the entire investment for interest and depreciation is represented by $25,000 yearly in this case, or 0.56 cent per kilowatt hour of probable output. Energy from this plant is transmitted and used in a large system of electrical supply.

If, through lack of water or inability to store water or energy at times when it is not wanted, generators cannot be operated at full capacity during the average number of hours assumed above, the item of interest and depreciation per unit of delivered energy must be higher than that computed. With the possible figure for this item at less than six-tenths of a cent per kilowatt hour, there is opportunity for some increase before it becomes prohibitive. At the plant last named the entire investment amounted to $166 per kilowatt capacity of connected generators, compared with $162 in the former case, and these figures may be taken as fairly representative for the development of water-power in a first-class manner on small rivers, under favorable conditions. In both of these instances the power-houses are quite close to the dams. If long canals or pipe lines must be built to convey the water, the expense of development may be greatly increased.

One advantage of water- over steam-power is the smaller cost of the building with the former for a given capacity of plant. The building for direct-connected electric generators, driven by water-wheels, is relatively small and simple. Space for fuel, boilers, economizers, feed-water heaters, condensers, steam piping, and pumps is not required where water-power is used. No chimney or apparatus for mechanical draught is needed.

The model electric station operated by water-power usually consists of a single room with no basement under it. One such station has floor dimensions 27 by 52 feet, giving an area of 1,404 square feet, and contains generators of 800 kilowatts capacity. This gives 1.75 square feet of floor space per kilowatt of generators. In this station there is ample room for all purposes, including erection or removal of machinery.

Next to the saving of fuel, the greatest advantage of water-power is due to the relatively small requirements for labor at generating stations where it is used. This is well illustrated by an example from actual practice. In a modern water-power station that contributes to electrical supply in a large city the generator capacity is 1,200 kilowatts. All of the labor connected with the operation of this station during nearly twenty-four hours per day is done by two attendants working alternate shifts.

These attendants live close to the station in a house owned by the electric company, and receive $60 each per month in addition to house rent. Considering the location, $12 per month is probably ample allowance for the rent. This brings the total expense of operation at this station for labor up to $132 per month, or $1,584 per year, a sum corresponding to $1.32 yearly per kilowatt of generator capacity.

At steam-power stations of about the above capacity, operating twenty-four hours daily, $6 is an approximate yearly cost of labor per kilowatt of generators in use. It thus appears that water-power plants may be operated at less than one-fourth of the labor expense necessary at steam stations per unit of capacity. On an average, the combined cost of fuel and labor at electric stations driven by steam-power is a little more than 76 per cent of their total cost of operation. Of this total, labor represents about 28, and fuel about 48 per cent. Water-power, by dispensing with fuel and with three-fourths of the labor charge, reduces the expense of operation at electric stations by fully 69 per cent.

But this great saving in the operating expenses of electric stations can be made only where water entirely displaces coal. If part water-power and part coal are used, the result depends on the proportion of each, and is obviously much affected by the variations of water-power capacity. In such a mixed system the saving effected by water-power must also depend on the extent to which its energy can be absorbed at all hours the day. By far the greater number of electric stations using water-power are obliged also to employ steam during either some months in the year or some hours in the day, or both.

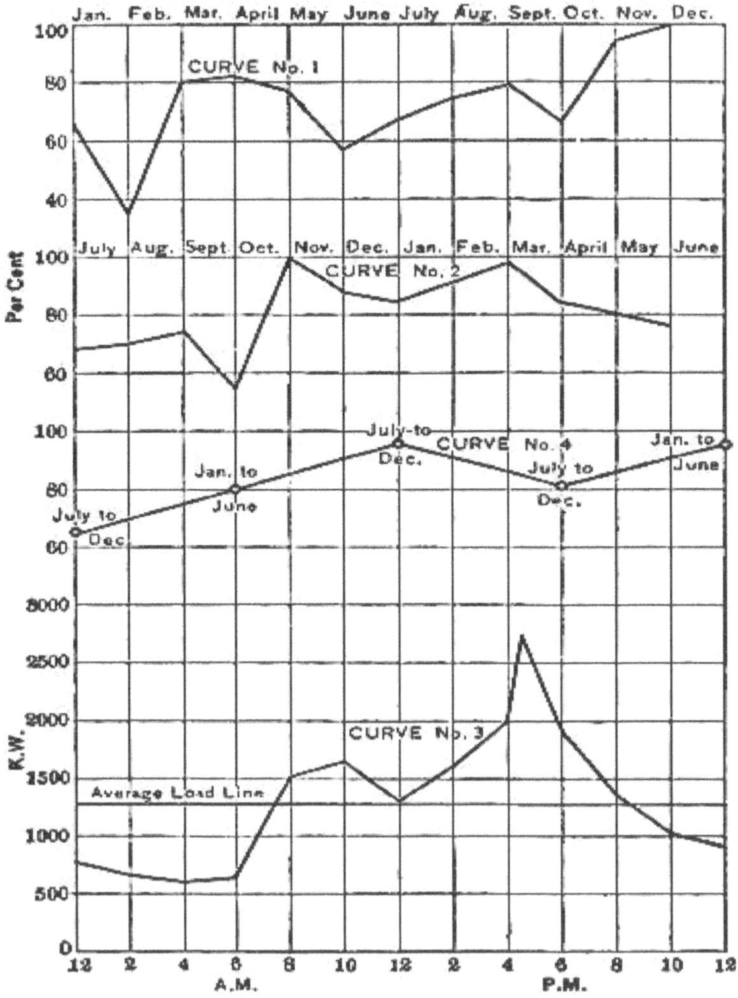

ENERGY CURVES FROM WATER POWER ELECTRIC STATIONS.
FIG. 3.

It is highly important, therefore, to determine, as nearly as may be, the answers to three questions:

First, what variations are to be expected in the capacity of a water-power during the several months of a year?

Second, if the daily flow of water is equal in capacity to the daily output of electrical energy, how far can the water-power be devoted to the development of that energy?

Third, with a water-power sufficient to carry all electrical loads at times of moderately high water, what percentage of the yearly output of energy in a general supply system can be derived from the water?

To the first of these questions experience alone can furnish an answer. Variations in the discharge of rivers during the different months of a year are very great. In a plant laid out with good engineering skill some provision will be made for the storage of water, and the capacity of generating equipment will correspond to some point between the highest and lowest rates of discharge.

Curve No. 1 in the diagram on the opposite page represents the energy output at an electric station driven entirely by water-power from a small stream during the twelve months of 1901, the entire flow of the stream being utilized. During December, 1901, the output of this station was 527,700 kilowatts, and was greater than that in any other month of the year. Taking this output at 100 per cent, the curve is platted to show the percentage attained by the delivered energy in each of the other months. At the lowest point on the curve, corresponding to the month of February, the output of energy was only slightly over 33 per cent of that in December. During nine other months of the year the proportion of energy output to that in December was over 60 and in three months over 80 per cent. For the twelve months the average delivery of energy per month was 73.7 per cent of that during December.

PERCENTAGES OF ENERGY DELIVERED
IN DIFFERENT MONTHS, 1901.

January	68.0
February	33.1
March	80.5
April	81.7
May	77.9
June	58.6
July	67.7
August	75.8
September	79.3
October	65.9
November	95.8

December 100.0

At a somewhat small water-power station on another river with a watershed less precipitous than that of the stream just considered, the following results were obtained during the twelve months ending June 30th, 1900. For this plant the largest monthly output of energy was in November, and this output is taken at 100 per cent. The smallest delivery of energy was in October, when the percentage was 53.1 of the amount for November. In each of seven other months of the year the output of energy was above 80 per cent of that in November. During March, April, May, and June the water-power yielded all of the energy required in the electrical supply system with which it was connected, and could, no doubt, have done more work if necessary. For the twelve months the average delivery of energy per month was 80.6 per cent of that in November, the month of greatest output.

PERCENTAGES OF ENERGY DELIVERED IN DIFFERENT MONTHS, 1899 AND 1900.

July	68.6
August	69.1
September	73.3
October	53.1
November	100.0
December	87.0
January	84.9
February	91.3
March	98.5
April	85.7
May	80.8
June	74.9

The gentler slopes and better storage facilities of this second river show their effect in an average monthly delivery of energy 6.9 per cent higher as to the output in a month when it was greatest than the like percentage for the water-power first considered. These two water-power illustrate what can be done with only very moderate storage capacities on the rivers involved. At both stations much water escapes over the dams during several months of each year. With enough storage space to retain all waters of these rivers until wanted the energy outputs could be largely increased.

As may be seen by inspection of curve No. 2, the second water-power has smaller fluctuations of capacity, as well as a higher average percentage of the maximum output than the water-power illustrated by curve No. 1.

If the discharge of a stream during each twenty-four hours is just sufficient to develop the electrical energy required in a supply system during that time, the water may be made to do all of the electrical work in one of two ways. If the water-power has enough storage capacity behind it to hold the excess of water during some hours of the day, then it is only necessary to install enough water-wheels and electric generators to carry the maximum load. Should the storage capacity for water be lacking, or the equipment of generating apparatus be insufficient to work at the maximum rate demanded by the electrical system, then an electric storage battery must be employed if all of the water is to be utilized and made to do the electrical work.

The greatest fluctuations between maximum and minimum daily loads at electric lighting stations usually occur in December and January. The extent of these fluctuations is illustrated by curve No. 3, which represents the total load on a large electrical supply system during a typical week-day of January, 1901. On this day the maximum load was 2,720 and the minimum load 612 kilowatts, or 22.5 per cent of the highest rate of output. During the day in question the total delivery of energy for the twenty-four hours was 30,249 kilowatt hours, so that the average load per hour was 1,260 kilowatts. This average is 46 per cent of the maximum load.

Computation of the area included by curve No. 3 above the average load line of 1,260 kilowatts shows that about 17.8 per cent of the total output of energy for the day was delivered above the average load, that is, in addition to an output at average load. It further appears by inspection of this load curve that this delivery of energy above the average load line took place during 12.3 hours of the day, so that its average rate of delivery per hour was 438 kilowatts.

If a water-power competent to carry a load of 1,260 kilowatts twenty-four hours per day be applied to the system illustrated by curve No. 3, then about 17.8 per cent of the energy of the water for the entire day must be stored during 11.7 hours and liberated in the remaining 12.3 hours. This percentage of the total daily energy of the water amounts to 36 per cent of its energy during the hours that storage takes place.

If all of the storage is done with water, the electric generators must be able to work at the rate of 2,720 kilowatts, the maximum load. If all of the storage is done in electric batteries, the use of water may be uniform throughout the day, and the generator capacity must be enough above 1,260 kilowatts to make up for losses in the batteries. Where batteries are employed the amount of water will be somewhat greater than that necessary to operate the load directly with generators, because of the battery losses.

In spite of the large fluctuations of electrical loads throughout each twenty-four hours, it is thus comparatively easy to operate them with water-powers that are little, if any, above the requirements of the average loads.

Perhaps the most important question relating to the use of water-power in electrical supply is what percentage of the yearly output of energy can be derived from water where this power is sufficient to carry the entire load during a part of the year. With storage area for all surplus water in any season, the amount of work that could be done by a stream might be calculated directly from the records of its annual discharge of water. As such storage areas for surplus water have seldom, or never, been made available in connection with electrical systems, the best assurance as to the percentage of yearly output that may be derived from water-power is found in the experience of existing plants.

The question now to be considered differs materially from that involving merely the variations of water-power in the several months, or even the possible yearly output from water-power. The ratio of output from water-power to the total yearly output of an electrical system includes the result of load fluctuations in every twenty-four hours and the variable demands for electrical energy in different months, as well as changes in the amount of water-power available through the seasons.

In order to show the combined result of these three important factors curve No. 4 has been constructed. This indicates the percentages of total semi-yearly outputs of electrical energy derived from water-power in two supply systems. Each half-year extends either from January to June, inclusive, or from July to December, inclusive, and thus covers a wet and dry season. Each half-year also includes a period of maximum and one of minimum demand for electrical energy in lighting. The period of largest water supply usually nearly coincides with that of heaviest lighting load, but this is not always true.

Electrical systems have purposely been selected in which the water-power in at least one month of each half-year was nearly or quite sufficient to carry the entire electrical load. The percentage of energy from water-power to the total energy delivered by the system is presented for each of five half-years. Three of the half-years each run from July to December, and two extend from January to June, respectively. The half years that show percentages of 66.8, 80.2, and 95.6, respectively, for the relation of energy from water-power to the total electrical output relate to one system, and the half years that show percentages of 81.97 and 94.3 for the energy from water-power relate to another system.

For the half-year when 66.8 per cent. of the output of the electrical system was derived from water-power, the total output of the system was 3,966,026 kilowatt hours. During the month of December in this half-year more than 98 per cent of the electrical energy delivered by the system was from water-power, though the average for the six months was only 66.8 per cent from water.

In the following six months, from January to June, the electrical supply system delivered 4,161,754 kilowatt hours, and of this amount the water-power furnished 80.2 per cent. For the six months just named, one month, May, saw 99 per cent of all the delivered energy derived from water-power.

The same system during the next half-year, from July to December, without any addition to its water-power development or equipment, got 95.6 per cent of its entire energy output from water-power, and this output amounted to 4,415,945 kilowatt hours. In one month of the half-year just named only 0.2 per cent of the output was generated with steam-power.

These three successive half years illustrate the fluctuations of the ratio between water-power outputs and the demands for energy on a single system of electrical supply. The percentage of 81.9 for energy derived from water-power during the half-year from July to December represents the ratio of output from water to the total for an electrical supply system where water generated 94 per cent of all the energy delivered in one month.

In the same system during the following six months, with exactly the same water-power equipment, the percentage of output from water-power was 94.3 of the total kilowatt-hours delivered by the system. This result was reached in spite of the fact that the total outputs of the system in the two half-years were equal to within less than one per cent.

The lesson from the record of these five half-years is that comparatively large variations are to be expected in the percentage of energy developed by water-power to the total output of electrical supply systems in different half-years. But, in spite of these variations, the portion of electrical loads that may be carried by water-power is sufficient to warrant its rapidly extending application to lighting and power in cities and towns.

CHAPTER III.
COST OF CONDUCTORS FOR ELECTRIC-POWER TRANSMISSION.

Electrical transmission of energy involves problems quite distinct from its development. A great water-power, or a location where fuel is cheap, may offer opportunity to generate electrical energy at an exceptionally low cost. This energy may be used so close to the point of its development that the cost of transmission is too small for separate consideration.

An example of conditions where the important problems of transmission are absent exists in the numerous factories grouped about the great water-power plants at Niagara and drawing electrical energy from it. In such a case energy flows directly from the dynamos, driven by water-power, to the lamps, motors, chemical vats, and electric heaters of consumers through the medium, perhaps, of local transformers. Here the costs and losses of transmitting or distributing equipments are minor matters, compared with the development of the energy.

If, now, energy from the water-power is to be transmitted over a distance of many miles, a new set of costs is to be met. In the first place, it will be necessary to raise the voltage of the transmitted energy much above the pressure at the dynamos in order to save in the weight and cost of conductors for the transmission line. This increase of voltage requires transformers with capacity equal to the maximum rate at which energy is to be delivered to the line. These transformers will add to the cost of the energy that they deliver in two ways: by the absorption of some energy to form heat, and by the sum of annual interest, maintenance, and depreciation charges on the price paid for them. Other additions to the cost of energy delivered by the transmission line must be made to cover the annual interest, maintenance, and depreciation charges on the amount of the line investment, and to pay for the energy changed to heat in the line.

Near the points where the energy is to be used, the transmission line must end in transformers to reduce the voltage to a safe figure for local distribution. This second set of transformers will further add to the cost of the delivered energy in the same ways as the former set.

From these facts it is evident that, to warrant an electrical transmission, the value of energy at the point of distribution should at least equal the value at the generating plant plus the cost of the transmission. Knowing the cost of energy at one end of the transmission line and its value at the other, the difference between these two represents the maximum cost at which the transmission will pay.

Three main factors are concerned in the cost of electric power transmission, namely, the transformers, the pole line, and the wire or conductors. These factors enter into the cost of transmitted energy in very different degrees, according to the circumstances of each case. The maximum and average rates of energy transmission, the total voltage, the percentage of line loss, and the length of the line mainly determine the relative importance of the transformers, pole line, and conductors in the total cost of delivered energy.

First cost of transformers varies directly with the maximum rate of transmission, and is nearly independent of the voltage, the length of the transmission, and the percentage of line loss. A pole line changes in first cost with the length of the transmission, but is nearly independent of the other factors. Line conductors, for a fixed maximum percentage of loss, vary in first cost directly with the square of the length of the transmission and with the rate of the transmission; but their first cost decreases as the percentage of line loss increases and as the square of the voltage of transmission increases.

If a given amount of power is to be transmitted, at a certain percentage of loss in the line and at a fixed voltage, over distances of 50, 100, and 200 miles, respectively, the foregoing principles lead to the following conclusions: The capacity of transformers, being fixed by the rate of transmission, will be the same for either distance, and their cost is therefore constant. Transformer losses, interest, depreciation, and repairs are also constant. The cost of pole line, depending on its length, will be twice as great at 100 and four times as great at 200 as at 50 miles. Interest, depreciation, and repairs will also go up directly with the length of the pole lines.

Line conductors will cost four times as much for the 100- as for the 50-mile transmission, because their weight will be four times as great, and the annual interest and depreciation will go up at the same rate. For the transmission of 200 miles the cost of line conductors and their weight will be sixteen times as great as the cost at 50 miles. It follows that interest, depreciation, and maintenance will be increased sixteen times with the 200-mile transmission over what they were at 50 miles, if voltage and line loss are constant.

A concrete example of the cost of electric power transmission over a given distance will illustrate the practical application of these principles. Let the problem be to deliver electrical energy in a city distant 100 miles from the generating plant. Transformers with approximately twice the capacity corresponding to the maximum rate of transmission must be provided, because one set is required at the generating and another at the delivery station. The cost of these transformers will be approximately $7.50 per horse-power for any large capacity.

Reliability is of the utmost importance in a great power transmission, and this requires a pole line of the most substantial construction. Such a line in a locality where wooden poles can be had at a moderate price will cost, with conductors in position, about $700 per mile, exclusive of the cost of the conductors themselves or of the right of way but including the cost of erecting the conductors. The 100 miles of pole line in the present case should, therefore, be set down at a cost of $70,000.

A large delivery of power must be made to warrant the construction of so long and expensive a line, and 10,000 horse-power may be taken as the maximum rate of delivery. On the basis of two horse-power of transformer capacity for each horse-power of the maximum delivery rate, transformers with a capacity of 20,000 horse-power are necessary for the present transmission. At $7.50 per horse-power capacity, the first cost of these transformers is $150,000.

Before the weight and cost of line conductors can be determined, the voltage at which the transmission shall be carried out and the percentage of the energy to be lost in the conductors at periods of maximum load must be decided on. The voltage to be used is a matter of engineering judgment, based in large part on experience, and cannot be determined by calculation. In a transmission of 100 miles the cost of conductors is certain to be a very heavy item, and, as this cost decreases as the square of the voltage goes up, it is desirable to push the voltage as high as the requirements for reliable service permit.

A transmission line 142 miles long, from the mountains to Oakland, Cal., has been in constant and successful use for several years with 40,000 volts pressure. This line passes through wet as well as dry climate. It seems safer to conclude, therefore, that 40,000 volts may be used in most places with good results.

Having decided on the amount of power and the voltage and length of the transmission, the required weight of conductors will vary inversely as the percentage of energy lost as heat in the line. The best percentage of loss depends on the number of factors, some of which, such as the cost of energy at the generating plant, are peculiar to each case.

As a provisional figure, based in part on the practice elsewhere, the loss on the line here considered may be taken at 10 per cent. when transmitting the full load of 10,000 horse-power. If the line is constructed on this basis the percentage of loss will be proportionately less for any smaller load. Thus, when the line is transmitting only 5,000 horse-power, the loss will amount to 5 per cent. During the greater portion of each day the demand for power is certain to be less than the maximum figure, so that a maximum loss of 10 per cent will correspond to an average loss on all the power delivered to the line of probably less than 7 per cent.

In order to deliver 10,000 horse-power by the transformers at a receiving station from a generating plant 100 miles distant where the pressure is 40,000 volts, the copper conductors must have a weight of about 1,500,000 pounds, if the loss of energy in them is 10 per cent of the energy delivered to the line. Taking these conductors at a medium price of 15 cents per pound, their cost amounts to $225,000.

The combined cost of the transformers, pole line, and line conductors, as now estimated, amounts to $445,000. No account is taken of the right-of-way for the pole line, because in many cases this would cost nothing, the public roads being used for the purpose; in other cases the cost might vary greatly with local conditions.

The efficiency of the transmission is measured by the ratio of the energy delivered by the transformers at the receiving station for local distribution to the energy delivered by the generating plant to the transformers that supply energy to the line for transmission. If worked at full capacity the large transformers here considered would have an efficiency of nearly 98 per cent; but as they must work, to some extent, on partial loads, the actual efficiency will hardly exceed 96 per cent.

The efficiency of the line conductors rises on partial loads, and may be safely taken at 93 per cent for all of the energy transmitted, though it is only 90 per cent on the maximum load. The combined efficiencies of the two sets of transformers and the line give the efficiency of the transmission, which equals the product of $0.96 \times 0.93 \times 0.96$, or almost exactly 85.7 per cent. In other words, the transformers at the water-power station absorb 1.17 times as much energy as the transformers at the receiving station deliver to distribution lines in the place of use.

Interest, maintenance, and depreciation of this complete transmission system are sufficiently provided for by an allowance of 15 per cent yearly on its entire first cost. As the total first cost of the transmission system was found to be $445,000, the annual expense of interest, depreciation, and repairs at 15 per cent of this sum amounts to $66,750.

In order to find the bearings of this annual charge on the cost of power transmission the total amount of energy transmitted annually must be determined. The 10,000 horse-power delivered by the system at the sub-station is simply the maximum rate at which energy may be supplied, and the element of time must be introduced in order to compute the amount of transmitted energy. If the system could be kept at work during twenty-four hours a day at full capacity, the delivered energy would be represented by the product of the numbers which stand for the capacity and for the total number of hours yearly.

Unfortunately, however, the demands for electric light and power vary through a wide range in the course of each twenty-four hours, and the period of maximum demand extends over only a small part of each day. The problem is, therefore, to find what relation the average load that may be had during the twenty-four hours bears to the capacity required to carry this maximum load. As the answer to this question depends on the power requirements of various classes of consumers, it can be obtained only by experience. It has been found that some electric stations, working twenty-four hours daily on mixed loads of lamps and stationary motors, can deliver energy to an amount represented by the necessary maximum capacity during about 3,000 hours per year. Applying this rule to the present case, the transformers at the sub-station, if loaded to their maximum capacity of 10,000 horse-power by the heaviest demands of consumers, may be expected to deliver energy to the amount of 3,000 × 10,000 = 30,000,000 horse-power hours yearly.

The total cost of operation for this transmission system was found above to be $66,750 per annum, exclusive of the cost of energy at the generating plant. This sum, divided by 30,000,000, shows the cost of energy transmission to be 0.222 cent per horse-power hour, exclusive of the first cost of the energy. To obtain the total cost of transmission, the figures just given must be increased by the value of the energy lost in transformers and in the line conductors. In order to find this value, the cost of energy at the generating plant must be known.

The cost of electrical energy at the switchboard in a water-power station is subject to wide variations, owing to the different investments necessary in the hydraulic work per unit of power developed. With large powers, such as are here considered, a horse-power hour of electrical energy may be developed for materially less than 0.5 cent in some plants. As the average efficiency of the present transmission has been found to be 85.7 per cent of the energy delivered by the generators, it is evident that 1.17 horse-power hours must be drawn from the generators for every horse-power hour supplied by the transformers at the sub-station for distribution. In other words, 0.17 horse-power hour is wasted for each horse-power hour delivered.

The cost of 0.17 of a horse-power hour, or say not more than $0.5 \times 0.17 = 0.085$ cent, must thus be added to the figures for transmission cost already found, that is, 0.222 cent per horse-power hour, to obtain the total cost of transmission. The sum of these two items of cost amounts to 0.307 cent per horse-power hour, as the entire transmission expense.

It may now be asked how the cost of transmission just found will increase if the distance be extended. As an illustration, assume the length of the transmission to be 150 instead of 100 miles. Let the amount of energy delivered by the sub-station, the loss in line conductors, and the energy drawn from the generating plant remain the same as before. Evidently the cost of the pole line will be increased 50 per cent, that is, from $70,000 to $105,000. Transformers, having the same capacity, will not be changed from the previous estimate of $150,000. If the voltage of the transmission remain constant, as well as the line loss at maximum load, the weight and cost of copper conductors must increase with the square of the distances of transmission. For 150 miles the weight of copper will thus be 2.25 times the weight required for the 100-mile transmission.

Instead of an increase in the weight of conductors a higher voltage may be adopted. The transformers for the two great transmission systems that extend over a distance of about 150 miles, from the Sierra Nevada Mountains to San Francisco Bay, in California, are designed to deliver energy to the line at either 40,000 or 60,000 volts, as desired. Though the regular operation at first was at the lower pressure, the voltage has been raised to 60,000.

The lower valleys of the Sacramento and the San Joaquin rivers, which are crossed by these California systems, as well as the shores of San Francisco Bay, have as much annual precipitation and as moist an atmosphere as most parts of the United States and Canada. Therefore there seems to be no good reason to prevent the use of 60,000 volts elsewhere.

The distance over which energy may be transmitted at a given rate, with a fixed percentage of loss and a constant weight of copper, goes up directly with the voltage employed. This rule follows because, while the weight of conductors to transmit energy at a given rate, with a certain percentage of loss and constant voltage, increases as the square of the distance, the weight of conductors decreases as the square of the voltage when all the other factors are constant.

Applying these principles to the 150-mile transmission, it is evident that an increase of the voltage to 60,000 will allow the weight of conductors to remain exactly where it was for the transmission of 100 miles, the rate of working and the line loss being equal for the two cases.

The only additional item of expense in the 150-mile transmission, on the basis of 60,000 volts, is the $35,000 for pole line. Allowing 15 per cent on the $35,000 to cover interest, depreciation, and maintenance, as before, makes a total yearly increase in the costs of transmission of $5,250 over that found for the transmission of 100 miles. This last sum amounts to 0.0175 cent per horse-power hour of the delivered energy.

The cost of transmission is thus raised to 0.307 + 0.0175 = 0.324 cent per horse-power hour of delivered energy on the 150-mile system with 60,000 volts.

Existing transmission lines not only illustrate the relations of the factors named above to the cost and weight of conductors, but also show marked variations of practice, corresponding to the opinions of different engineers. In order to bring out the facts on these points, the data of a number of transmission lines are here presented. On these lines the distance of transmission varies between 5 and 142 miles, the voltage from 5,000 to 50,000, and the maximum rate of work from a few hundred to some thousands of horse-power. For each transmission the single length and total weight of conductors, the voltage, and the capacity of the generating equipment that supplies the line is recorded. From these data the volts per mile of line, weight and cost of conductors per kilowatt capacity of generating equipment, and the weight of conductors per mile for each kilowatt of capacity in the generating equipment are calculated. In each case the length of line given is the distance from the generating to the receiving station. The capacity given for generating equipment in each case is that of the main dynamos, where their entire output goes to the transmission line in question, but where the dynamos supply energy for other purposes also, the rating of the transformers that feed only the particular transmission line is given as the capacity of generating equipment.

Distance and Voltage of Electrical Transmission.

	Distance in Miles.		Volts.	Volts per Mile.
Colgate to Oakland, Cal.	142		60,000	422
Cañon Ferry to Butte, Mont.	65		50,000	769
Santa Ana River to Los Angeles	83		33,000	397
Ogden to Salt Lake City, Utah	36	.5	16,000	438
Madrid to Bland, N. M.	32		20,000	625
Welland Canal to Hamilton, Can.	35-37		22,500	643
San Gabriel Cañon to Los Angeles	23		16,000	695
Cañon City to Cripple Creek, Colo.	23	.5	20,000	851
Apple River to St. Paul, Minn.	25		25,000	1,000
Yadkin River to Salem, N. C.	14	.5	12,000	827
Into Victor, Colo.	8		12,600	1,575
Montmorency Falls to Quebec	7		5,500	785
Farmington River to Hartford	11		10,000	909
Sewall's Falls to R.R. shops Concord	5	.5	10,000	1,818
Wilbraham to Ludlow Mills	4	.5	11,500	2,555
To Dales, Ore.	27		22,000	814

The transmission systems here considered have been selected because it was possible to obtain the desired data as to each, and it may be presumed that they fairly illustrate present practice. It may be noted at once that in general the line voltage is increased with the length of the transmission. Thus, the transmission for the Ludlow Mills over a distance of 4.5 miles is carried out at 11,500 volts. On the other hand, the transmission between Cañon Ferry and Butte, a distance of 65 miles, employs 50,000 volts and represents recent practice. The system from Colgate to Oakland, a distance of 142 miles, the longest here considered, now has 60,000 volts on its lines. In spite of the general resort to high pressures with greater distances of transmission, the rise in voltage has not kept pace with the increasing length of line. For the

Wilbraham-Ludlow transmission the total pressure amounts to 2,555 volts per mile, while the line from Colgate to Oakland with 31.5 times the length of the former operates at an average of only 422 volts per mile. Of the fifteen transmissions considered, six are over distances of less than 15 miles, and for four of the six the voltage is more than 900 per mile. Eight transmissions range from 23 to 83 miles in length, with voltages that average between 1,000 volts per mile at 25 miles and only 397 per mile on the 83-mile line. The volts per mile are 6 times as great in the Ludlow as in the Oakland transmission.

CAPACITY OF GENERATING STATIONS AND WEIGHT OF CONDUCTORS.

Location of Transmission.	Kilowatt Capacity at Generators.	Total Weight of Conductors.	Pounds of Conductors per Kilowatt Capacity.
Wilbraham to Ludlow	4,600	17,820	3.7 [A]
Sewall's Falls to railroad shops	50	6,914	15
Into Victor, Colo.	1,600	15,960	10
To Dales, Ore.	1,000	33,939	34
Apple River to St. Paul	3,000	159,600	53
Farmington River to Hartford	1,500	54,054	36
Cañon City to Cripple Creek	1,500	59,079	39
Yadkin River to Salem	1,500	58,073	39
Montmorency Falls to Quebec	2,400	189,056	79
Cañon Ferry to Butte	5,700	658,320	115
San Gabriel Cañon to Los Angeles	1,200	73,002	61
Welland Canal to Hamilton	6,000	376,494	63
Madrid to Bland, N. M.	600	127,680	212
Ogden to Salt Lake City	2,250	292,365	129
Santa Ana River to Los Angeles	2,250	664,830	295
Colgate to Oakland	11,250	906,954	81
		446,627	40 [A]

[A] Aluminum.

These wide variations in the volts per mile on transmission lines and in length of lines lead to different weights of conductors per kilowatt of generator capacity. All other factors remaining constant, the weight of conductors per kilowatt of generator capacity would be the same whatever the length of the transmission, provided that the volts per mile were uniform for all cases. One important factor, the percentage of loss for which the line conductors are designed at full load, is sure to vary in different cases, and lead to corresponding variations in the weights of conductors per kilowatt of generator capacity. In conductors of equal length one pound of aluminum has nearly the same electrical resistance as two pounds of copper, and this ratio must be allowed for when copper and aluminum lines are compared.

From the table it may be seen that the weight of conductors per kilowatt of generator capacity for the transmission from Santa Ana River is 29.5 times as great as the like weight for the line into Victor. But the volts per mile are four times as great on the Victor as they are on the Santa Ana River line. The extreme range of the cases presented is that between the Ludlow plant, with the equivalent of 7.4 pounds, and the Santa Ana River system with 295 pounds of copper conductors per kilowatt of generator capacity. Three transmissions with 1,575 to 2,555 volts per mile have the equivalent of 7.4 to 15 pounds of copper each, per kilowatt of generator capacity.

WEIGHT AND COST OF CONDUCTORS.

	Pounds per Kilowatt Mile.		Dollars per Generator Kilowatt.
Wilbraham to Ludlow	0	.86[A]	1.11
Sewall's Falls to railroad shops	2	.7	2.25
Into Victor, Colo.	0	.9	1.50
To Dales, Ore.	1	.2	5.10
Apple River to St. Paul	2	.1	7.95
Farmington River to Hartford	3	.2	10.80
Cañon City to Cripple Creek	1	.6	5.85
Yadkin River to Salem	2	.6	5.85
Montmorency Falls to Quebec	11	.2	11.85
Cañon Ferry to Butte	1	.7	17.25
San Gabriel Cañon to Los Angeles	2	.6	9.85

Welland Canal to Hamilton	1	.7	9.45
Madrid to Bland, N. M.	6	.6	31.80
Ogden to Salt Lake City	3	.5	19.35
Santa Ana River to Los Angeles	3	.5	44.25
Colgate to Oakland	-	.56 .27[A]	24.15

[A] Aluminum.

Of the seven transmissions using between 36 and 79 pounds of copper for each kilowatt of generator capacity, four have voltages ranging from 827 to 1,000 per mile, and on only one is the pressure as low as 643 volts per mile. Five transmission lines vary between 115 and 295 pounds of copper, or its equivalent, per kilowatt of generator capacity, and their voltages per mile are as high as 769 in one case and down to 281 in another. Allowing for some variations in the percentages of loss in transmission lines at full load, the fifteen plants plainly illustrate the advantage of a high voltage per mile, as to the weight of conductors. This advantage is especially clear if the differences due to the lengths of the transmissions are eliminated by dividing the weight of conductors per kilowatt of generator capacity in each case by the length of the transmission in miles. This division gives the weight of conductors per kilowatt of generators for each mile of the line, which may be called the weight per kilowatt mile. For the Ludlow transmission this weight is only 0.86 pound of aluminum, the equivalent of 1.72 pounds of copper, while the like weight for the line into Quebec is 11.2 pounds of copper, or 6.5 times that for the former line. But the voltage per mile on the Ludlow is 3.2 times as great as the like voltage on the Quebec line.

The weight of conductor per kilowatt mile in the Victor line is only 0.9 pound, and the like weight for the line between Madrid and Bland is 6.6 pounds, or 7.3 times as great. On the Victor line the voltage per mile is 2.5 times as great as the voltage for each mile of the Bland line.

Comparing systems with nearly equal voltages per mile, it appears in most cases that only such difference exists in their pounds of conductors per kilowatt mile as may readily be accounted for by designs for various percentages of loss at full load. Though the transmission line into Butte is nearly twice as long as the one entering Hamilton, the weight of conductors for each is 1.7 pounds per kilowatt mile. The line from Santa Ana River is more than twice as long as the one entering Salt Lake City, but its voltage per mile is only nine per cent less, and there are 3.5 pounds of copper in each line per kilowatt mile.

The final, practical questions as to conductors in electrical transmission relate to their cost per kilowatt of maximum working capacity, and per kilowatt hour of delivered energy. If the cost of conductors per kilowatt of generator capacity is greater than that of all the remaining equipment, it is doubtful whether the transmission will pay. If fixed charges on the conductors more than offset the difference in the cost of energy per kilowatt hour at the points of development and delivery, it is certain that the generating plant should be located where the power is wanted. The great cost of conductors is often put forward as a most serious impediment to long-distance transmission, and the examples here cited will indicate the weight of this argument. In order to find the approximate cost of conductors per kilowatt of generator capacity for each of the transmission lines here considered, the price of bare copper wire is taken at 15 cents, and the price of bare aluminum wire at 30 cents per pound. In each case the weight of copper or aluminum conductor per kilowatt of generator capacity is used to determine their costs per kilowatt of this capacity at the prices just named. This process when carried out for the 15 transmission lines shows that their cost of conductors per kilowatt of generator capacity varies between $1.11 for the 4.5 mile line into Ludlow and $44.25 for the line of 83 miles from the Santa Ana River. It should be noted that the former of these lines operates at 2,555 and the latter at 397 volts per mile. The line into Madrid shows an investment in conductors of $31.80 per kilowatt of generator capacity with 625 volts per mile. That a long transmission does not necessarily require a large investment in conductors per kilowatt of generator capacity is shown by the line 65 miles long into Butte, for which the cost is $17.25 per kilowatt, with 769 volts per mile. For the transmission to St. Paul, a distance of 25 miles, at 1,000 volts per mile, the cost of conductors is $7.95 per kilowatt of generator capacity. The seven-mile line into Quebec shows an investment of $11.85 per kilowatt of generator capacity.

CHAPTER IV.
ADVANTAGES OF THE CONTINUOUS AND ALTERNATING CURRENT.

Electrical transmissions over long distances in America have been mainly carried out with alternating current. In Europe, on the other hand, continuous current is widely used on long transmissions at high voltages. So radical a difference in practice seems to indicate that neither system is lacking in points of superiority.

A fundamental feature of long transmissions is the high voltage necessary for economy in conductors, and this voltage is attained by entirely different methods with continuous and alternating currents. In dynamos of several hundred or more kilowatts capacity the pressure of continuous current has not thus far been pushed above 4,000 volts, because of the danger of sparking and flashing at the commutator. Where 10,000 or more volts are required on a transmission line with continuous current a number of dynamos are connected in series so that the voltage of each is added to that of the others. In this way the voltage of each dynamo may be as low as is thought desirable without limiting the total line voltage. There is no apparent limit to the number of continuous-current dynamos that may be operated in series or to the voltage that may be thus obtained. In the recently completed transmission from St. Maurice to Lausanne, Switzerland, with continuous current, ten dynamos are connected in series to secure the line voltage of 23,000. When occasion requires twenty or thirty or more dynamos to be operated in series, giving 50,000 or 75,000 volts on the line, machines exactly like those in the transmission just named, may be used. No matter how many of these dynamos are operated in series the electric strain on the insulation of the windings of each dynamo remains practically constant, because the iron frame of each dynamo is insulated in a most substantial manner from the ground. The electric strain on the insulation of the windings of each dynamo in the series is thus limited to the voltage generated by that dynamo. There is no practical limit to the thickness or strength of the insulation that may be interposed between the frame of each dynamo and the ground, and hence no limit to line voltage as far as dynamo insulation is concerned.

It is impracticable to operate alternating dynamos in series so as to add their voltages, and the pressure available in transmission with alternating current must be that of a single dynamo or must be obtained by the use of transformers. The voltage of an alternating may be carried much higher than that of a continuous-current dynamo of very large capacity, and in many cases pressures of 13,200 volts are now supplied to transmission lines by alternating dynamos. Just how high the voltage of single alternating dynamos will be carried no one can say, but it seems probable that the practical limit will prove to be much less than the voltages now employed in some transmissions. As the voltage of alternating dynamos is carried higher the thickness of insulation on their armature coils and consequently the size or number of slots in their armature cores and the size of these cores increase rapidly. The dimensions and weight of an alternating dynamo per unit of its capacity thus go up with the voltage, and at some undetermined point the cost of the high-voltage dynamo is greater than that of a low-voltage dynamo of equal capacity with raising transformers. To the voltage that may be supplied by transformers there is no practical limit now in sight. Lines have been in regular operation from one to several years on which transformers supply 40,000 to 50,000 volts; some large transformers have been built for commercial use at 60,000 volts, and other transformers for experimental and testing purposes have been employed in a number of cases for pressures of 100,000 volts and more.

Available voltages for continuous- and alternating-current transmissions are thus on a practically equal footing as to their upper limit. The amount of power that may be generated and delivered with either the alternating- or continuous-current system of transmission is practically unlimited. Single alternating dynamos may be had of 5,000 or even 8,000 kilowatts capacity if desired, but it is seldom that these very large units are employed, because the capacity of a generating station should be divided up among a number of machines. It is perhaps impracticable to build single continuous-current dynamos with capacities equal to those of the largest alternators, but as any number of the continuous-current machines may be operated either in series or multiple, the power that may be applied to a transmission circuit is unlimited.

At the plant or plants where the power transmitted by continuous current is received, a number of motors must be connected in series to operate at the high-line voltage. These motors may all be located in a single room, may be connected to machinery in different parts of a building, or may be in use at points miles apart. The vital requirement is that the motors must be in series with each other so that the line voltage divides between them. If simply mechanical power is wanted at the places where the motors are located, they complete the transmission system and no further electrical apparatus is required. Where, however, as at Lausanne, the transmitted power is to be used in a system of general electrical supply, the motors that receive the current at the line voltage must drive dynamos that will deliver energy of the required sorts. In the station at Lausanne four of the motors to which the transmission line is connected each drives a 3,000-volt three-phase alternator for the distribution of light and power. The fifth motor at this station drives a 600-volt dynamo which delivers continuous current to a street railway. A sixth motor in the same series drives a cement factory some distance from the station. Neglecting minor changes in capacity due to losses in the line and motors, this continuous-current system must thus include three kilowatts in motors and dynamos for each kilowatt delivered for general electrical distribution at the receiving station. In a case in which only mechanical power is wanted at the receiving station, the dynamos and motors concerned in the transmission must have a combined capacity of two horse-power for each horse-power delivered at the motor shaft. In contrast with these figures, the electrical equipment in a transmission with alternating current for mechanical power alone includes two kilowatts capacity in generators and motors, besides two kilowatts capacity in transformers for each corresponding unit of power delivered at the motor shaft unless generators and motors operate at the full line voltage. If a general electrical supply is to be operated by the alternating system of transmission, either motors and dynamos or rotary converters must be added to transformers where continuous current is required. An alternating transmission may thus include as little as one kilowatt in dynamos and one in transformers, or as much as two kilowatts capacity in dynamos, two in transformers, and one in motors for each kilowatt delivered to distribution lines at the receiving station.

Line construction from the continuous-current transmission is of the most simple character apart from the necessity of high insulation. Only two wires are necessary and they may be of any desired cross-section, strung on a single pole line and need not be transposed. On these wires the maximum voltage for which insulation must be provided is the nominal voltage of the system. It is possible under these conditions to build a single transmission line with two conductors of such size and strength and at such a distance apart that a high degree of reliability is attained against breaks in the wires or arcing between them. In a transmission of power by two- or three-phase alternating current at least three wires are necessary and six or more are often employed. If six or more wires carrying current at the high voltages required by long transmissions are mounted on a single line of poles, it is not practicable to obtain such distances between the wires as are desirable. The repair of one set of wires while the other set is in operation is a dangerous task, and an arc originating between one set of the wires is apt to be communicated to another set. For these reasons two pole lines are frequently provided for a transmission with alternating current, and three or more wires are then erected on each line. Compared with a continuous-current transmission, one with alternating current often requires more poles and is quite certain to require more cross-arms, pins, insulators, and labor of erection. For a given effective voltage of transmission it is harder to insulate an alternating- than a continuous-current line. In the first place the maximum voltage of the alternating line with even a true sine curve of pressure is 1.4

times the nominal effective voltage, but the insulation must withstand the maximum pressure. Then comes the matter of resonance, which may carry the maximum voltage of an alternating circuit up to several times its normal amount, if the period of electrical vibration for that particular circuit should correspond to the frequency of the dynamos that operate it. Even where the vibration period of a transmission circuit and the frequency of its dynamos do not correspond, and good construction should always be planned for this lack of agreement, resonance may and often does increase the normal voltage of an alternating transmission by a large percentage. The alternating system of transmission must work at practically constant voltage whatever the state of its load, so that the normal stress on the insulation is always at its maximum. In a transmission with continuous current on the other hand, if the prevailing practice of a constant current and varying pressure on the line is followed, the insulation is subject to the highest voltage only at times of maximum load on the system. Lightning is a very real and pressing danger to machinery connected to long transmission lines, and this danger is much harder to guard against in an alternating system than in a system with continuous constant current. The large degree of exemption from damage by lightning enjoyed by series arc dynamos is well known, the magnet windings of such machines acting as an inductance that tends to keep lightning out of them. Moreover, with any continuous-current machines lightning arresters having large self-induction may be connected in circuit and form a most effective safeguard against lightning, but this plan is not practicable on alternating lines.

In the matter of switches, controlling apparatus, and switchboards, an alternating transmission requires much more equipment than a system using continuous, constant current. The ten dynamos in the generating station at St. Maurice, with a capacity of 3,450 kilowatts at 23,000 volts, are each connected and disconnected with the transmission by a switch in a small circular column of cast-iron that stands hardly breast high. An amperemetre and voltmetre are mounted on each dynamo. The alternating generators in a station of equal capacity and voltage would require a large switchboard fitted with bus-bars, oil switches, and automatic circuit-breakers. Relative efficiencies for the continuous-current and the alternating-transmission systems vary with the kind of service required at receiving stations and with the extent to which transformers are used in the alternating system, other factors being constant. For purposes of comparison the efficiency at full load of both alternating- and continuous-current dynamos and motors, also of rotary converters, may be fairly taken at 92 per cent, and the efficiency of transformers at 96 per cent.

For the line an efficiency of 94 per cent may be assumed at full load, this being the actual figure in one of the Swiss transmissions of 2,160 kilowatts at 14,400 volts to a distance of 32 miles. Where the continuous current system must simply deliver mechanical power at the receiving stations, its efficiency under full load amounts to 92 × .94 × .92 = 79.65 per cent from dynamo shaft to motor shaft. An alternating system delivering mechanical power will have an efficiency of 92 × .94 × .96 × .92 = 76.46 per cent between dynamo shaft and motor shaft, if the line voltage is generated in the armature coils of the dynamo and the line loss is 6 per cent. If step-up transformers are employed to secure the line voltage the efficiency of the alternating transmission delivering mechanical power drops to the figure of 92 × .96 × .94 × .96 × .92 = 73.40 per cent. It thus appears that for the simple delivery of mechanical power the continuous current transmission has an advantage over the alternating of three to six per cent in efficiency, depending on whether step-up transformers are employed.

When the receiving station must deliver a supply of either continuous or alternating current for general distribution, the efficiency of the continuous-current transmission amounts to 92 × .94 × .92 × .92 = 73.27 per cent. The alternating-transmission system in a case in which no step-up transformers are employed will deliver alternating current of the same frequency as that on the transmission line at any desired pressure for general distribution at an efficiency of 92 × .94 × .96 = 83.02 per cent, if step-down transformers are used, but the efficiency drops to 83.02 × .96 = 79.70 per cent. when step-up transformers are introduced. If the alternating transmission uses no step-up transformers and delivers either alternating or continuous current by means of motor generators, its efficiency at full load is 83.02 × .92 × .92 = 70.26 per cent, but with step-up transformers added the efficiency drops to 70.26 × .96 = 67.43 per cent. In a transmission where electrical energy must be delivered for general distribution, the full-load efficiency of an alternating system ranges either higher or lower than that of a continuous-current system depending on whether the current from the transmission line must be converted or not.

Line loss is the same whatever the load in a constant-current transmission, so that line efficiency falls rather rapidly with the load. On the other hand, at constant pressure the percentage of energy loss on the line varies directly with the load, but the actual rate of energy loss with the square of the load. On partial loads the line efficiency is thus much higher with alternating than with continuous constant current.

Efficiency of electrical machinery is generally low at partial loads, so that in cases in which the number or capacity of alternating dynamos, transformers, motors, or rotary converters for a transmission would be greater per unit of delivered power than the corresponding number or capacity of machines for a transmission by continuous current, the latter would probably have the advantage in the combined efficiency of machinery at partial loads. In this way the lower-line efficiency of one system might offset the lower efficiency of machinery in the other. Energy is usually very cheap at the generating station of a transmission system. For this reason small differences in the efficiencies of different systems should be given only moderate weight in comparison with the items of first cost, reliability, and expense of operation.

In the matter of first cost at least the continuous-current system seems to have a distinct advantage over the alternating. Without going into a detailed estimate, it is instructive to consider the figures given by a body of five engineers selected to report on the cost of continuous- and alternating-current equipments for the St. Maurice and Lausanne transmission. According to the report of these engineers, a three-phase transmission system would have cost $140,000 more than the continuous-current system actually installed, all other factors remaining constant. It should be noted that the conditions of this transmission are favorable to three-phase working and unfavorable to continuous-current equipment, because all of the energy except that going to the 400 horse-power motor at the cement mill must be delivered at the receiving station for general distribution. Moreover, four out of the five motors at Lausanne drive three-phase generators, and only one drives a continuous-current dynamo for the electric railway, so that a three-phase transmission would have required only one rotary converter. Had the transmission been concerned merely with the delivery of mechanical power, as at the cement mill, the advantage of the continuous- over the alternating-current system in the matter of first cost would have been much greater than it was.

Long-distance transmission with three-phase current began at Frankfort, in 1891, when 58 kilowatts were received over a 25,000-volt line from Lauffen, 109 miles away. Shortly after this historic experiment, three-phase transmission in the United States began on a commercial scale, and plants of this sort have multiplied rapidly here. Meantime very little has been done in America with continuous currents in long transmissions. In Europe, the birthplace of the three-phase system, it has failed to displace continuous current for transmission work. About a score of these continuous-current transmissions are already at work there. If the opinion of European engineers as to the lower cost of the continuous-current system, all other factors being equal, is confirmed by experience, this current will yet find important applications to long transmissions in the United States.

Systems of transmission with continuous-current may operate at constant voltage and variable current, at constant current and variable voltage, or with variations of both volts and amperes to correspond with changes of load. Dynamos of several thousand kilowatts capacity each can readily be had at voltages of 500 to 600, but the attempt to construct dynamos to deliver more than two or three hundred kilowatts each at several thousand volts has encountered serious sparking at the commutator. Thus far, dynamos that yield between 300 and 400 kilowatts each have been made to give satisfactory results at pressures as high as 2,500 volts.

Another one of the Swiss transmissions takes place over a distance of thirty-two miles at 14,400 volts, the capacity being 2,160 kilowatts. To give this voltage and capacity, eight dynamos are connected in series at the generating station, each dynamo having an output of 150 amperes at 1,800 volts, or 216 kilowatts.

Continuous-current motors are, of course, subject to the same limitations as dynamos in the matter of capacity at high voltage, so that a series of motors must be employed to receive the high-pressure energy from the line. The number of these motors may just equal, or may be less or greater than the number of dynamos, but the total working voltage of all the motors in operation at one time must equal the total voltage of the dynamos in operation at that time minus the volts of drop in the line.

Each constant-current motor may have any desired capacity up to the practicable maximum, but it must be designed for the current of the system. The voltage at the terminals of each motor varies with its load, being greatest when the motor is doing the most work. Constant speed is usually attained at each motor by means of a variable resistance connected across the terminals of the magnet coils. The amount of this resistance is regulated by a centrifugal governor, driven by the motor shaft. This governor also shifts the position of the brushes on the commutator to prevent sparking as the current flowing through the magnet coils is changed.

For a constant-current transmission the magnet and armature windings of both dynamos and motors are usually connected in series with each other and the line so that the same current passes through every element of the circuit, except that each motor may have some current shunted out of its magnet coil for the purpose of speed regulation.

In some cases, however, the magnet coils of the dynamos are connected in multiple with each other and receive their current from a separate dynamo designed for the purpose. With this separate excitation of the magnet coils, the dynamo armatures are still connected in series with each other and the line.

The total voltage at the generating station and on the line of a constant-current system varies with the rate at which energy is delivered, and has its maximum value only at times of full load. To obtain this variation of voltage, it is the general practice to change the speed of the dynamos by means of an automatic regulator which is actuated by the line current. Any increase of the line current actuates the regulator and reduces the speed of the dynamos, while a decrease of the line current raises the dynamo speed. With a good regulator the variations of the line current are only slight. Under this method of regulation the dynamos in operation have a substantially constant current in both armature and magnet coils at all times, so that there is no reason to shift the position of the brushes on the commutator.

Generating stations of constant current transmission systems are generally driven by water-power and the speed regulator operates to change the amount of water admitted to each wheel. Each turbine wheel usually drives a pair of dynamos, but one or any number of dynamos might be driven by a single wheel. The two dynamos driven by a single wheel are generally connected in series at all times, and are cut in or out of the main circuit together. When the load on a constant-current generating station is such that the voltage can be developed by less than all the dynamos, one or more dynamos may be stopped and taken out of the circuit.

To do this the dynamo or pair of dynamos to be put out of service may be stopped, their magnet coils having first been short-circuited, and then a switch across the connections of their armatures to the lines closed, after which the connections of the armatures to the line are opened. By a reverse process, any dynamo or pair of dynamos may be cut into the operating circuit.

At the terminals of each dynamo in the series, while in operation, the voltage is simply that developed in its armature, so that the insulation between the several windings is subject to only a corresponding stress. The entire voltage of the line, however, tends to force a current from the coils of the dynamo at one end of the series into its frame, thence to any substance on which that frame rests, and so on to the frame and coils of the dynamo at the other end of the series. To protect the insulation of the dynamo coils from the line voltage, thick blocks of porcelain are placed beneath the dynamo frames, and the armature shafts are connected to those of the turbines by insulating couplings.

Besides the switches, already mentioned, a voltmeter and ammeter should be provided for each dynamo and also for the entire series of machines. This completes the switchboard equipment, which is, therefore, very simple. As the line loss of a constant-current system is the same whatever the load that is being operated, this loss may be a large percentage of the total output when the load is light. If, for illustration, five per cent of the maximum voltage of the station is required to force the constant current through the line, the percentage of line loss will rise to ten when the station voltage is one-half the maximum, and to twenty when the station is delivering only one-quarter of its full capacity.

In view of this property of constant-current working, the line loss should be made quite small in its ratio to the maximum load, as most stations must work on partial loads much of the time. Five per cent of maximum station voltage is a fair general figure for the line loss in a constant-current transmission, but the circumstances of a particular case may dictate a higher or a lower percentage.

On the 32-mile transmission, above named, the loss in the line is six per cent of the station output at full load.

If a transmission with continuous current is to be carried out at constant pressure the limitation as to the capacity and voltage of each dynamo is about the same as with constant current. Probably more energy is now transmitted by continuous current at constant pressure than by any other method, the greater part being devoted to electric railway work at 500 to 600 volts. Dynamos for about these voltages can readily be had in capacities up to several thousand kilowatts each, but the length of transmission that can be economically carried out at this pressure is comparatively small. For each kilowatt delivered to a line at 500 volts and to be transmitted to a distance of five miles at a ten per cent loss in the line, the weight of copper conductors must be 372 pounds, costing $56.80 at 15 cents per pound. This sum is twice to four times the cost of good continuous-current dynamos per kilowatt of capacity. If the distance of transmission is ten miles and the voltage and line loss remain as before, the weight of copper conductor must be increased to 1,488 pounds per kilowatt delivered to the line, costing $227.20.

Experience has shown that in sizes of not more than 400 kilowatts, continuous-current dynamos may safely have a voltage of 2,000 each, and any number of such dynamos may be operated in multiple, giving whatever capacity is desired. At 2,000 volts and a loss of 10 per cent in the line the weight of copper conductors per kilowatt would be 93 pounds, costing $13.95, for each kilowatt delivered to the line on a 10-mile transmission. With 2,000 volts on a 20-mile transmission the weight of conductors per kilowatt would be the same as their weight on a 5-mile transmission at 500 volts, the percentage of loss being equal in the two cases. Large continuous-current motors of, say, 50 kilowatts or more can be had for a pressure of 2,000 volts, so that any number of such motors might be operated from a 2,000-volt, constant-pressure line entirely independent of each other. From these figures it is evident that a transmission of 10 miles may be carried out with continuous-current at constant pressure from a single dynamo with good efficiency and a moderate investment in conductors.

When the distance is such that much more than 2,000 volts are required for the constant-pressure transmission, with continuous current, resort must be had to the connection of dynamos and motors in series. Any number of dynamos may be so connected as in the case of constant-current work. The combined voltages of the series of motors connected to the constant-pressure transmission line must equal the voltage of that line, so that the number of motors in any one series must be constant. If the voltage of transmission is so high that more than two or three motors must be connected in each series, there comes the objection that motors must be operated at light loads during much of the time. Moreover, each series of motors must be mechanically connected to the same work, as that of driving a single dynamo or other machine, because if the loads on the motors of a series vary differently, these motors will not operate at constant speed. Continuous-current transmission at constant pressure with motors in series thus lacks the flexibility of transmission at constant current where any motor may be started and stopped without regard to the others in the series, the line voltage being automatically regulated at the generating station according to the number of motors in use at any time and to the work they are doing.

In the efficiency of its dynamos, motors and line, a constant-pressure system of transmission is substantially equal to one with constant current at full load. At partial loads the constant-pressure line has the advantage because the loss of energy in it varies with the square of the load. Thus at constant pressure the line loss in energy per hour at half-load is only one-fourth as great as the loss at full load. On the other hand, the energy loss in the constant-current line is the same at all stages of load. Because of these facts it is good practice to allow, say, a ten-per-cent loss in a constant-pressure line and only five per cent in a constant-current line at full load.

In a generating station at 2,000 volts or more constant pressure, it is desirable to have the magnet coils of the main dynamos connected in multiple and separately excited by a small dynamo at constant pressure. This plan is especially desirable when the armatures of several dynamos are connected in series to obtain the line voltage. Separately excited magnet coils make it easier to control the operation of the several dynamos, coils of low-voltage are cheaper to make than coils of high voltage, and the low voltage windings are less liable to burn out. If a series of constant-pressure motors is in use at one point, it may be cheaper and safer to excite its magnet coils from a special dynamo than from the line.

In a transmission carried out with series-wound dynamos and motors, the speed of the motors may be constant at all loads without any special regulating mechanism. To attain this result it is necessary that all the motors be coupled so as to form a single unit mechanically and that the dynamos be driven at constant speed. A transmission system of this sort may include a single dynamo and a single motor, or two or more dynamos, and two or more motors may be used in series.

When the dynamos of such a system are driven at constant speed and a variable load is applied to the single motor, or to the mechanically connected motors, both the voltage of the system and the amperes flowing in all its parts change together so that practically constant speed is maintained at the motors, provided that the design of both the dynamos and motors is suitable for the purpose. With the maximum load on the motors the volts and amperes of the system have their greatest values, and these values both decline with smaller loads. The chief disadvantage of this system lies in the fact that where more than one motor is employed all the motors must be mechanically joined together so as to work on the same load.

Compared with the constant-current system, this combination of series dynamos with mechanically connected series motors has the distinct advantage that neither the dynamos nor motors require any sort of regulators in order to maintain constant motor speed. It is only necessary that the dynamos be driven at constant speed and that both the dynamos and motors be designed for the transmission. In comparison with a constant-pressure system, the one under consideration has the advantage that neither its dynamos nor motors require magnet coils with a high voltage at their terminals and composed of fine wire or separate excitation by a special dynamo. These features of the system with series dynamos and motors, the latter being joined as a mechanical unit, make it cheaper to install and easier to operate than either of the other two. This system is especially adapted for the delivery of mechanical power in rather large units. The voltage available may be anything desired, but is subject to the practical limitations that all the motors must deliver their power as a mechanical unit, so that unless the power is quite large the number of motors in the series and, therefore, the voltage is limited.

An interesting illustration of the system of transmission just described exists between a point on the River Suze, near Bienne, Switzerland, and the Biberest paper mills. At the river a 400 horse-power turbine water-wheel drives a pair of series-wound dynamos, each rated at 130 kilowatts and 3,300 volts. These dynamos are connected in series, giving a total capacity of 260 kilowatts and a pressure of 6,600 volts. At the Biberest mills are located two series-wound motors, mechanically coupled and connected in series with each other and with the two-wire transmission line, which extends from the two dynamos at the River Suze. Each of these motors has a capacity and voltage equal to that of either of the dynamos previously mentioned. The coupled motors operate at the constant speed of 200 revolutions per minute at all loads and deliver over 300 horse-power when doing maximum work. Between the generating plant at the river and the Biberest mills the distance is about 19 miles, and the two line wires are each of copper, 275 mils, or a little more than one-fourth inch in diameter. The dynamos and motors of this system are mounted on thick porcelain blocks in order to protect the insulation of their windings from the strain of the full-line voltage.

Either of the three systems of transmission by continuous-current that have been considered requires a smaller total capacity of electrical apparatus for a given rate of mechanical power delivery than any system using alternating current except that where both the dynamos and motors operate at line voltage.

CHAPTER V.
THE PHYSICAL LIMITS OF ELECTRIC-POWER TRANSMISSION.

Electrical energy may be transmitted around the world if the line voltage is unlimited. This follows from the law that a given power may be transmitted to any distance with constant efficiency and a fixed weight of conductors, provided the voltage is increased directly with the distance.

The physical limits of electric-power transmission are thus fixed by the practicable voltage that may be employed. The effects of the voltage of transmission must be met in the apparatus at generating and receiving stations on the one hand, and along the line on the other. In both situations experience is the main guide, and theory has little that is reliable to offer as to the limit beyond which the voltage will prove unworkable.

Electric generators are the points in a transmission system where the limit of practical voltage is first reached. In almost all high-voltage transmissions of the present day in the United States alternating generators are employed. Very few if any continuous-current dynamos with capacities in the hundreds of kilowatts and voltages above 4,000 have been built in Europe, and probably none in the United States. Where a transmission at high voltage is to be accomplished with continuous current, two or more dynamos are usually joined in series at the generating station, and a similar arrangement with motors is made at the receiving station, so that the desired voltage is available at the line though not present at any one machine.

Alternating dynamos that deliver current at about 6,000 volts have been in regular use for some years, in capacities of hundreds of kilowatts each, and may readily be had of several thousand kilowatts capacity. But even 6,000 volts is not an economical pressure for transmissions over fifteen to fifty miles, such as are now quite common; consequently in such transmissions it has been the rule to employ alternators that operate at less than 3,000 volts, and to raise this voltage to the desired line pressure by step-up transformers at the generating station. More recently, however, the voltage of alternating generators has been pushed as high as 13,000 in the revolving-magnet type where all the armature windings are stationary. This voltage makes it practicable to dispense with the use of step-up transformers for transmissions up to or even beyond 30 miles in some cases. This voltage of 13,000 in the armature coils is attained only by constructions involving some difficulty because of the relatively large amount of room necessary for the insulating materials on coils that develop this pressure. The tendency of this construction is to give alternators unusually large dimensions per given capacity. It seems probable, moreover, that the pressures developed in the armature coils of alternating generators must reach their higher limits at a point much below the 50,000 and 60,000 volts in actual use on present transmission lines. In the longest transmissions with alternating current there is, therefore, little prospect that step-up transformers at the generating stations and step-down transformers at receiving stations can be dispensed with. The highest voltage that may be received or delivered at these stations is simply the highest that it is practicable to develop by transformers and to transmit by the line.

A very high degree of insulation is much more easy to attain in transformers than in generator armatures, because the space that can be readily made available for insulating materials is far greater in the transformers, and further because their construction permits the complete immersion of their coils in petroleum. This oil offers a much greater resistance than air to the passage of electric sparks, which tend to set up arcs between coils at very high voltages and thus destroy the insulation. Danger to insulation from the effect known as creeping between coils at widely different pressures is largely avoided by immersion of the coils in oil. For several years groups of transformers have been worked regularly at 40,000 to 60,000 volts, and in no instance is there any indication that the upper limit of practicable voltage has been reached. On the contrary, transformers have repeatedly been worked experimentally up to and above 100,000 volts.

From all these facts, and others of similar import, it is fair to conclude that the physical limit to the voltages that it is practicable to obtain with transformers is much above the 50,000 or 60,000 volts now in practical use on transmission systems. So far as present practice is concerned, the limit to the use of high voltages must be sought beyond the transformers and outside of generating and receiving stations. As now constructed, the line is that part of the transmission system where a physical limit to the use of higher voltages will first be reached. The factors that tend most directly to this limit are two: temporary arcing between the several wires on a pole, and the less imposing but constant passage of energy from one wire to another. On lines of very high voltage arcing is occasionally set up by one of several causes. At a point where one or more of the insulators on which the wires are mounted become broken or defective, the current is apt to flow from one wire to another along a wet cross-arm, until the wood grows carbonized and an arc is formed that ends by burning up the cross-arm or even the pole. Where lines are exposed to heavy sea fog, the salt is in some cases deposited on the insulators and cross-arms to an extent that starts an arc between the wires, and ends often in the destruction of the cross-arm. In some instances the glass and porcelain insulators supporting wires used with high voltages are punctured by sparks that pass right through the material of the insulator to the pin on which it is mounted, thus burning the pin and ultimately the cross-arm. This trouble is easily met, however, by the adoption of a better grade of porcelain or of an insulator with a greater thickness of glass or porcelain between the wires and the supporting pin. Arcs between lines at high voltages usually start by sparks that jump from the lower edges of insulators, when they are wet or covered with salt deposit, to the cross-arm. As the lower edges of insulators are only a few inches from their cross-arms, the sparks find a path of comparatively low resistance by passing from insulator to cross-arm and thence to the other insulator and wire. The wood of a wet cross-arm is a far better conductor than the air. Where wires are several feet or more apart, sparks probably never jump directly through the air from one to the other. Large birds flying close to such wires, however, have in some instances started momentary arcs between them. The treatment of cross-arms with oil or paraffine reduces the number of arcs that occur on a line of high voltage, but does not do away with them.

As the voltages of long transmissions have gone up, the distance through the air between wires and the distances between the lower wet edges of insulators and the cross-arms have been much increased. Most of the earlier transmission lines for high voltages were erected on insulators spaced from one to two feet apart. In contrast with this practice, the three wires of the transmission line in operation at 50,000 volts between Cañon Ferry and Butte are arranged at the corners of a triangle seventy-eight inches apart, one wire at the top of each pole and the other two at opposite ends of the cross-arm. A voltage that would just start an arc along a wet cross-arm between wires eighteen inches apart would be quite powerless to do so over seventy-eight inches of cross-arm, the lower wet edges of insulators being equidistant from cross-arms in the two cases. To reach the cross-arm, the electric current passes down over the wet or dirty outside surface of the insulator to its lower edge. In the older types of insulators the lower wet edge often came within two inches of the cross-arm. For the 50,000-volt line just mentioned the insulators (see illus.) are mounted with their lower wet edges about eight inches above the cross-arms. At its lower edge each insulator has a diameter of nine inches, and a small glass sleeve extends several inches below this edge and close to the wooden pin, to prevent sparking from the lower wet edge of the insulator to the pin. These increased distances between wires in a direct line through the air, and also the greater distances between the lower wet edges of insulators and their pins and cross-arms, are proving fairly effective to prevent serious arcing under good climatic conditions, for the maximum pressures of 50,000 to 60,000 volts now in use. If these voltages are to be greatly exceeded it is practically certain that the distance between wires, and from the lower wet edges of insulators to the wood of poles or cross-arms, must be still further increased to avoid destructive arcing.

The nearest approach to an absolute physical limit of voltage with present line construction is met in the constant current of energy through the air from wire to wire of a circuit. A paper in vol. XV., Transactions American Institute Electrical Engineers, gives the tests made at Telluride, Col., to determine the rates at which energy is lost by passing through the air from one wire to another of the same circuit. The tests at Telluride were made with two-wire circuits strung on a pole line 11,720 feet in length, at first with iron wires of 0.165 inch diameter and then with copper wires of 0.162 inch diameter. Measurements were made of the energy escaping from wire to wire at different voltages on the line, and also with the two wires at various distances apart. It was found that the loss of energy over the surfaces of insulators was very slight, and that the loss incident to the passage of energy directly through the air is the main one to be considered. This leakage through the air varies with the length of the line, as might be expected. Tests were made with pairs of wires running the entire length of the pole line and at distances of 15, 22, 35, and 52 inches apart respectively. Losses with wires 22 or 35 inches apart were intermediate to the losses when wires were 15 and 52 inches apart respectively. Results given in the original paper for the pair of wires that were 15 inches apart and for the pair that were 52 inches apart are here reduced to approximate watts per mile of two-wire line. At 40,000 volts the loss between the two wires that were 15 inches apart was about 150 watts per mile, and between the two wires that were 52 inches apart the loss was 84 watts per mile. The two wires 15 inches apart showed a leakage of approximately 413 watts per mile when the voltage was up to 44,000, but the wires 52 inches apart were subject to a leakage of only 94 watts per mile at the same voltage. At 47,300 volts, the highest pressure recorded for the two wires 15 inches apart, the leakage between them was about 1,215 watts per mile, while an equal voltage on the two wires 52 inches apart caused a leakage of only 122 watts per mile, or one-tenth of that between the wires that were 15 inches apart. When about 50,000 volts were reached on the two wires 52 inches apart, the leakage between them amounted to 140 watts per mile; but beyond this voltage the loss went up rapidly, and was 225 watts per mile at about 54,600 volts. For higher pressures the loss between these two wires still more rapidly increased, and amounted to 1,368 watts per mile with about 59,300 volts, the highest pressure recorded. With a loss of about 1,215 watts per mile between the two wires 52 inches apart, the voltage on them was 58,800, in contrast with the 47,300 volts producing the same leakage on the two wires 15 inches apart.

Evidently, however, at even 52 inches between line wires the limit of high voltage is not far away. When the voltage on the 52-inch line was raised from 54,600 to 59,300, the leakage loss between the two wires increased about 1,143 watts per mile. If the leakage increases at least in like proportion, as seems probable, for still higher pressures, the loss between the two wires would amount to 6,321 watts per mile with 80,000 volts on the line. On a line 200 miles long this loss by leakage between the two wires would amount to 1,264,200 watts. Any such leakage as this obviously sets an absolute, physical limit to the voltage, and consequently the length of transmission.

Fortunately for the future delivery of energy at great distances from its source, the means to avoid the limit just discussed are not difficult. Other experiments have shown that with a given voltage and distance between conductors the loss of energy from wire to wire decreases rapidly as their diameters increase. The electrical resistance of air, like that of any other substance, increases with the length of the circuit through it. The leakage described is a flow of electrical energy through the air from one wire to another of the same circuit. To reduce this leakage it is simply necessary to give the path from wire to wire through the air greater electrical resistance by increasing its length, that is, by placing the wires at greater distances apart. The fact demonstrated at Telluride, that with 47,300 volts on each line the leakage per mile between the two wires 15 inches apart was ten times as great as the leakage between the two wires 52 inches apart, is full of meaning. Evidently, leakage through the air may be reduced to any desired extent by suitable increase of distance between the wires of the same circuit. But to carry this increase of distance between wires very far involves radical changes in line construction. Thus far it has been the uniform practice to carry the two or three wires of a transmission circuit on a single line of poles, and in many cases several such circuits are mounted on the same pole line. For the 65 mile transmission into Butte, Mont., only the three wires of a single circuit are mounted on one pole line, and this represents the best present practice. The cross-arms on this line are each 8 feet long, and one is attached to each pole. The poles are not less than 35 feet long and have 8-inch tops. One wire is mounted at the top of each pole, and the other two wires near the ends of the cross-arm, so that the three wires are equidistant and 78 inches apart. By the use of still heavier poles the length of cross-arms may be increased to 12 or 14 feet, for which their section should be not less than 4 by 6 inches. Placing one wire at the pole top, the 12-foot cross-arm would permit the three wires of a circuit to be spaced about 10.5 feet apart. The cost of extra large poles goes up rapidly and there are alternative constructions that seem better suited to the case. Moreover, a few tens of thousands of volts above present practice would bring us again to a point where even 10.5 feet between wires would not prevent a prohibitive leakage. Two poles might be set 20 feet apart, with a cross-piece between them, extending out 5 feet beyond each pole and having a total length of 30 feet. This would permit three wires to be mounted along the cross-piece at points about 14 feet apart.

If the present transmission pressures of 50,000 to 60,000 volts are to be greatly exceeded, the line structure may involve the use of a separate pole for each wire of a circuit, each wire to be mounted at the top of its pole. This construction calls for three lines of poles to carry the three wires of a three-phase transmission. Each of these poles may be of only moderate dimensions, say 30 feet long with 6- or 7-inch top. The cost of three of these poles will exceed by only a moderate percentage that of a 35- or 40-foot pole with an 8- to 10-inch top, such as would be necessary with 12-foot cross-arms. The distance between these poles at right angles to the line may be anything desired, so that leakage from wire to wire through the air will be reduced to a trifling matter at any voltage. Extra long pins and insulators at the pole tops will easily give a distance of two feet or more between the lower wet edge of each insulator and the wood of pin or pole. Such line construction would probably safely carry two or three times the maximum voltage of present practice, and might force the physical limit of electrical transmission back to the highest pressure at which transformers could be operated. With not more than 60,000 volts on the line the size of conductors is great enough in many cases to keep the loss of energy between them within moderate limits when they are six feet apart, but with a large increase of voltage the size of conductors must go up or the distances between them must increase.

CHAPTER VI.
DEVELOPMENT OF WATER-POWER FOR ELECTRIC STATIONS.

Electrical transmission has reduced the cost of water-power development. Without transmission the power must be developed at a number of different points in order that there may be room enough for the buildings in which it is to be utilized. This condition necessitates relatively long canals to conduct the water to the several points where power is to be developed, and also a relatively large area of land with canal and river frontage.

With electrical transmission the power, however great, may well be developed at a single spot and on a very limited area of land. The canal in this case may be merely a short passageway from one end of a dam to a near-by power-house, or may disappear entirely when the power-house itself forms the dam, as is sometimes the case.

These differences between the distribution of water for power purposes and the development by water of electrical energy for transmission may be illustrated by many examples.

A typical case of the distribution of water to the points where power is wanted may be seen in the hydraulic development of the Amoskeag Manufacturing Company at Manchester, N. H. This development includes a dam across the Merrimac River, and two parallel canals that follow one of its banks for about 3,400 feet down stream. By means of a stone dam and a natural fall a little beyond its toe a water head of about forty-eight feet is obtained at the upper end of the high-level canal. Below this point there is little drop in the bed of the river through that part of its course that is paralleled by the two canals. All of the power might be thus developed within a few rods of one end of the dam, if means were provided for its distribution to the points where it must be used.

Years ago, when this water-power was developed, the electrical transmission or distribution of energy was unheard of, and distribution of the water itself had therefore to be adopted. For this purpose the two canals already mentioned were constructed along the high bank of the river at two different levels.

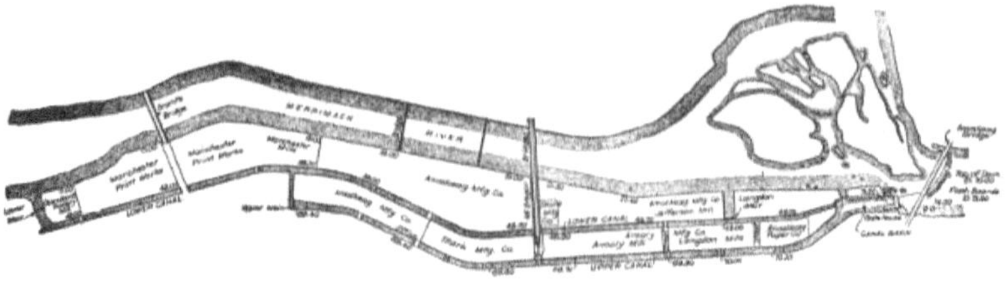

FIG. 4.—Hydraulic Development on the Merrimac River.

The high-level canal, so called, was designed to take water directly from the basin or forebay a little below one end of the dam, so that between this canal and the river there is a full water head of about forty-eight feet. Over nearly its entire course the nearer side of this high-level canal runs between 450 and 750 feet from the edge of the river wall, and thus includes between it and the river a large area on which factories to be driven by water-wheels may be located. It was thought, however, that this strip of land between the high-level canal and the river was too wide for a single row of mill sites, and the lower level canal was therefore constructed parallel with that on the higher level, but with about twenty-one feet less elevation.

Between these two canals a strip of land about 250 feet wide was left for the location of mills. By this arrangement of canals it is possible to supply wheels located between the high and the low levels with water under a head of about twenty-one feet, and to supply wheels between the lower canal and the river with water under a head of about twenty-nine feet. The entire area of land between the high canal and the river is thus made readily available for factory buildings.

Water for the lower canal is drawn mainly from the high canal through the wheels in buildings that are located between the two canals. It is desirable in a case of this sort to have as much water flow through the wheels between the high and low canal as flows through the wheels between the low canal and the river, but this is not always possible. A gate is therefore provided at the forebay where the two canals start, by which water may pass from the forebay directly into the low canal when necessary, but the head of twenty-one feet between the forebay and the low canal is lost as to this water. Between the high and low canal, and between the low canal and the river twenty-three turbine wheels or pairs of wheels have been connected, and these wheels have a combined rating of 9,500 horse-power.

To carry out this hydraulic development it thus appears that about 1.3 miles of canal have been constructed; one-half this length of river-front has been required, and about one-sixth square mile of territory has been occupied. Contrast with this result what might have been done if electrical transmission of power had been available at the time when this water-power was developed. All but a few rods in length of the existing 1.3 miles of canal might have been omitted, and an electric generating station with wheels to take the entire flow of the river might have been located not far from one end of the dam. Factories utilizing the electric power thus developed might have been located at any convenient points along the river-front or elsewhere, and no space would have been made unavailable because of the necessity of head- and tail-water connections to scattered sets of wheels.

Compare with the foregoing hydraulic development that at Cañon Ferry on the Missouri River, in Montana, where 10,000 horse-power is developed under a water-head of 32 feet. At Cañon Ferry the power-house is 225 feet by 50 feet at the floor level inside, contains turbine wheels direct-connected to ten main generators of 7,500 kilowatts, or 10,000 horse-power combined capacity, and is built on the river bank close to one end of the 500-foot dam. The canal which runs along the land side of the power-house, and takes water at the up-stream side of the abutment, is about twice the length of the power-house itself. The saving in the cost of canal construction alone, to say nothing of the saving as to the required area of land, is evidently a large item in favor of the electrical development and transmission. In its small area and short canal the Cañon Ferry plant is not an exception, but is rather typical of a large class of electric water-power plants that operate under moderate heads.

A like case may be seen in the plant at Red Bridge, on the Chicopee River, in Massachusetts, where a canal 340 feet long, together with penstocks 100 feet long, convey water from one end of the dam and deliver it to wheels in the electric station with a head of 49 feet. This station contains electric generators with a combined capacity of 4,800 kilowatts or 6,400 horse-power, and its floor area is 141 by 57 feet.

FIG. 5.—Canal at Red Bridge on Chicopee River.

So, again, at Great Falls, on the Presumpscot River, in North Gorham, Me. (see cut), the electric station sets about 40 feet in front of the forebay wall, which adjoins one abutment of the dam, and there is no canal whatever, as short penstocks bring water to the wheels with a head of 35 feet. In ground area this station is 67.5 by 55 feet, and its capacity in main generators is 2,000 kilowatts or 2,700 horse-power.

A striking illustration of the extent to which electrical transmission reduces the cost of water-power development may be seen at Gregg's Falls on the Piscataquog River, in New Hampshire, where an electric station of 1,200 kilowatts capacity has been built close to one end of the dam, and receives water for its wheels under a head of 51 feet through a short penstock, 10 feet in diameter, that pierces one of the abutments.

FIG. 6.

Perhaps the greatest electric water-power station anywhere that rests close to the dam that provides the head for its wheels is that at Spier Falls (see cut), on the upper Hudson. One end of this station is formed by the high wall section of the dam, and from this wall the length of the station down-stream is 392 feet, while its width is 70 feet 10 inches, both dimensions being taken inside. The canal or forebay in this case, like that at Cañon Ferry, lies on the bank side of the power-station, and is about equal to it in length. From this canal ten short penstocks, each 12 feet in diameter, will convey water under a head of 80 feet to as many sets of turbine wheels in the station. These wheels will drive ten generators with an aggregate capacity of 24,000 kilowatts or 32,000 horse-power.

Sometimes the slope in the bed of a river is so gradual or so divided up between the number of small falls, or the volume of water is so small, that no very large power can be developed at any one point without the construction of a long canal. In a case of this sort electrical transmission is again available to reduce the expense of construction that will make it possible to concentrate all the power from a long stretch of the river at a single point. This is done by locating electric generating stations at as many points as may be thought desirable along the river whose energy is to be utilized, and then transmitting power from all of these stations to the single point where it is wanted.

A case in point is that of Garvins Falls and Hooksett Falls on the Merrimac River and four miles apart. At the former of these two falls the head of water is twenty-eight feet, and at the latter it is sixteen feet. To unite the power of both these falls in a single water-driven station would obviously require a canal four miles long whose expense might well be prohibitive. Energy from both these falls is made available at a single sub-station in Manchester, N. H., by a generating plant at both points and transmission lines thence to that city.

At Hooksett the present capacity of the electric station is 1,000 horse-power, and at Garvins Falls the capacity is 1,700 horse-power. The river is capable of developing larger powers at both of these falls, however, and construction is now under way at Garvins that will raise its station capacity to 5,000 horse-power.

A similar result in the combination of water-powers without the aid of a long canal is reached in the case of Gregg's Falls and Kelley's Falls, which are three miles apart on the Piscataquog River. At the former of these two falls the electric generating capacity is 1,600 horse-power, as previously noted, and at the latter fall the capacity is 1,000 horse-power. In each case the station is close to its dam, and no canal is required. Electrical transmission unites these two powers in the same sub-station at Manchester that receives the energy from the two stations above named on the Merrimac River.

Instead of transmitting power from two or more waterfalls to some point distant from each of them, the power developed at one or more falls may be transmitted to the site of another and there used. This is, in fact, done at the extensive Ludlow twine mills on the Chicopee River, in Massachusetts. These mills are located at a point on the river where its fall makes about 2,500 horse-power available, and this fall has been developed to its full capacity. After a capacity of 2,400 horse-power in steam-engines had been added, more water-power was sought, and a new dam was located on the same river at a point about 4.5 miles up-stream from the mills. The entire flow of the river was available at this new dam, and a canal 4.5 miles long might have been employed to carry the water down to wheels at the mills in Ludlow.

Such a canal would have meant a large investment, not only for land and construction, but also, possibly, for damages to estates bordering on the river, if all of its water was diverted. Instead of such a canal, an electric generating station was located close to the new dam with a capacity of 6,400 horse-power, and this power is transmitted to motors in the mills at Ludlow.

Even where the power is to be utilized at some point distant from each of several waterfalls, it may be convenient to combine the power of all at one of them before transmitting it to the place of use. This is actually done in the case of two electric stations located respectively at Indian Orchard and Birchem Bend on the Chicopee River, whose energy is delivered to the sub-station of the electrical supply system in Springfield, Mass. At the Indian

Orchard station the head of water is 36 feet, and at Birchem Bend it is 14 feet, while the two stations are about 2 miles apart. A canal of this length might have been built to give a head of 50 feet at the site of the Birchem Bend dam, but instead of this an electric station was located near each fall, and a transmission line was built between the two stations. Each generating station was also connected with the sub-station in Springfield by an independent line, and power is now transmitted from one generating plant to the other, as desired, and the power of both may go to the sub-station over either line. In the Indian Orchard station the dynamo capacity is about 2,000 kilowatts, and at Birchem Bend it is 800 kilowatts.

Another case showing the union of two water-powers by electrical transmission, where the construction of an expensive canal was avoided, is that of the electrical supply system of Hartford, Conn. This system draws a large part of its energy from two electric plants on the Farmington River, at points that are about 3 miles apart in the towns of Windsor and East Granby, respectively. At one of these plants the head of water is 32 feet, and at the other it is 23 feet, so that head of 55 feet might have been obtained by building a canal 3 miles long. Each of these stations is located near its dam, and the generator capacity at one station is 1,200 and at the other 1,500 kilowatts. Transmission lines deliver power from both of these plants to the same sub-station in Hartford.

Sometimes two or more water-powers on the same river that are to be united are so far apart that any attempt to construct a canal between them would be impracticable. This is illustrated by the Spier and Mechanicsville Falls on the Hudson River, which are 25 miles apart in a direct line and at a greater distance by the course of the stream. At Spier Falls the head is 80 feet, and at Mechanicsville it is 18 feet. Union of the power of these two falls is thus out of the question for physical reasons alone. Electrical transmission, however, brings energy from both of these water-powers to the same sub-stations in Schenectady, Albany, and Troy.

In another class of cases electrical transmission does what could not be done by any system of canals, however expensive; that is, unites the water-powers of distinct and distant rivers at any desired point. Thus power from both the Merrimac and the Piscataquog rivers is distributed over the same wires in Manchester; the Yuba and the Mokelumne contribute to electrical supply along the streets of San Francisco; and the Monte Alto and Tlalnepantla yield energy in the City of Mexico.

It does not follow from the foregoing that it is always more economical to develop two or more smaller water-powers at different points along a river for transmission to some common centre than it is to concentrate the water at a single larger station by more elaborate hydraulic construction, and then deliver all of the energy over a single transmission line. The single larger hydraulic and electric plant will usually have a first cost larger than that of the several smaller ones, because of the required canals or pipe lines. A partial offset to this larger hydraulic investment is the difference in cost between one and several transmission lines, or at least the cost of the lines that would be necessary between the several smaller stations in order to combine their energy output before its transmission over a single line to the point of use.

Against the total excess of cost for the single larger hydraulic and electrical plant there should be set the greater expense of operation at several smaller and separate plants. Even a small water-driven electric station that can be operated by a single attendant at any one time must have two attendants if it is to deliver energy during the greater part or all of every twenty-four hours. But a single attendant can care for a water-power plant of 2,000 horse-power or more capacity, so that two plants of 750 horse-power each will require double the operating force of one plant of 1,500 horse-power. If two such plants are constructed instead of one that has their combined capacity, the monthly wages of the two additional operators will amount to at least one hundred dollars. If money is worth six per cent yearly, it follows that an additional investment of $\$1,200 \div 0.06 = \$20,000$ might be made in hydraulic work to concentrate the power at one point before the annual interest charge would equal the increase of wages made necessary by two plants.

Reliability of operation is one of the most important requirements in an electric water-power plant, and its construction should be carried out with this in view. Anchor ice is a serious menace to the regular operation of water-wheels in cold climates, because it clogs up the openings in the racks and in the wheel passages. Anchor ice is formed in small particles in the water of shallow and fast-flowing streams, and tends to form into masses on solid substances with which the water comes in contact.

At the entrance to penstocks or wheel chambers, steel racks with long, narrow openings, say one and one-quarter inches wide, are regularly placed to keep all floating objects away from the wheels. When water bearing fine anchor or frazil ice comes in contact with these racks, it rapidly clogs up the narrow openings between the bars, unless men are kept at work raking off the ice as it forms. At the Niagara Falls electric station, in some instances, when the racks become clogged, they have been lifted, and the anchor ice allowed to pass down through the wheels. This is said to have proved an effective remedy, but it would obviously be of no avail in a case where the ice clogged the passages of the wheels themselves.

The best safeguard against anchor ice is a large, deep forebay next to the racks, where the water, being comparatively quiet, will soon freeze over after cold weather commences. The anchor ice coming down to this forebay and losing most of its forward motion, soon rises to the surface or to the under side of the top coating of solid ice, and the warmer water sinks to the bottom. Good construction puts the entrance ends of penstocks well below the surface of water in the forebay, so that they may receive the warmer water that contains little or no anchor ice.

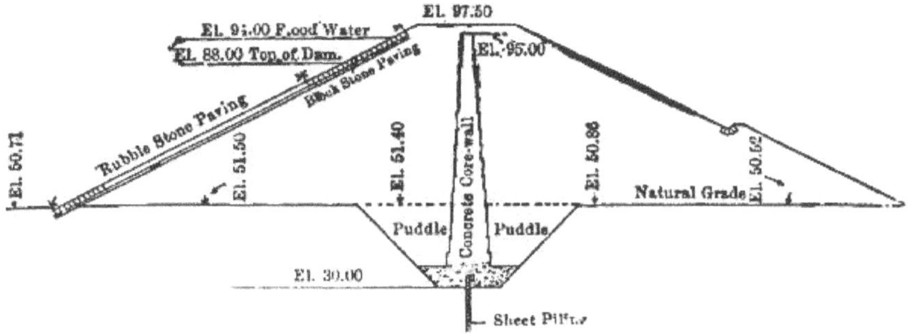

FIG. 7.—Cross Section of Dike on Chicopee River at Red Bridge.

Illustrations of practice along these lines, as to size, depth of forebay, and location of penstocks may be seen in many well-designed plants. One instance is that at Garvins Falls, on the Merrimac River, where the new hydraulic development for 5,000 horse-power is now under way. Water from the river in this case comes down to the power-station through a canal 500 feet long, and of 68 feet average width midway between the bottom and the normal flow line. In depth up to his flow line the canal is 12 feet at its upper and 13 feet at its lower end, just before it widens into the forebay. In this forebay the depth increases to 17 feet, and the width at the rack is double that of the canal. The steel penstocks, each 12 feet in diameter, terminate in the forebay wall at an average distance of 7 feet behind the rack, and each penstock has its centre 10.6 feet below the water level in the forebay. As there is a large pond created by the dam in this case, and as the flow of water in the canal is deep rather than swift, enough opportunity is probably afforded for any anchor ice to rise to the surface before it reaches the forebay in this case.

Penstocks for the electric station at Great Falls, on the Presumpscot River, whence energy is drawn for lighting and power in Portland, Me., are each 8 feet in diameter, and pierce the forebay wall behind the rack with their centres 15 feet below the normal water level in the forebay. In front of the forebay wall the water stands 27 feet deep, and the pond formed by the dam, of which the forebay wall forms one section, is 1,000 feet wide and very quiet. Though the Maine climate is very cold in winter and the Presumpscot is a turbulent stream above the dam and pond, there has never been any trouble with anchor ice at the Great Falls plant. An excellent illustration is thus presented of the fact that deep, still water in the forebay is a remedy for troubles with ice of this sort.

Maximum loads on electrical supply systems are usually from twice to four times as great as the average loads during each twenty-four hours. A pure lighting service tends toward the larger ratio between the average and maximum load, while a larger motor capacity along with the lamps, tends to reduce the ratio. Furthermore, by far the greater part of the energy output of an electrical supply system during each twenty-four hours must be delivered between noon and midnight. For these reasons there must be enough water stored, that can flow to the station as wanted, to carry a large share of the load during each day, unless storage batteries are employed to absorb energy at times of light load, if the entire normal flow of the river is to be utilized.

It is usually much cheaper to store water than electrical energy for the daily fluctuations of load, and the only practicable place for this storage is most commonly behind the dam that maintains the head for the power-station. This storage space should be so large that the drain upon it during the hours of heavy load will lower the head of water on the wheels but little, else it may be hard to maintain the standard speed of revolution for the wheels and generators, and consequently the transmission voltage.

FIG. 8.—Valley Flooded above Spier Falls on the Hudson River.

At the Great Falls plant, water storage to provide for the fluctuations of load in different parts of the day takes place back of the dam, and for about one mile up-stream. This dam is 450 feet long in its main part, and a retaining wall increases the total length to about 1,000 feet. For half a mile up-stream from this dike and dam the average width of the pond is 1,000 feet, and its greatest depth is not less than 27 feet. As the station capacity is 2,700 horse-power in main generators, with a head of 35 feet at the wheels the storage capacity is more than ample for all changes of load at different times of day.

The dam at Spier Falls, on the Hudson River, is 1,820 feet long between banks, 155 feet high above bedrock in its deepest section, and raises the river 50 feet above its former level. Behind the dam a lake is formed one-third of a mile wide and 5 miles long. Water from this storage reservoir passes down through the turbines with a head of 80 feet, and is to develop 32,000 horse-power. As a little calculation will show, this lake is ample to maintain the head under any fluctuation in the daily load. At Cañon Ferry, where electrical energy for Butte and Helena, Mont., is developed, the dam, which is 480 feet long, crosses the river in a narrow canyon that extends up-stream for about half a mile. Above this canyon the river valley widens out, and the dam, which maintains a head of 30 feet at the power-station, sets back the water in this valley, and thus forms a lake between two and three miles wide and about seven miles long. At the station the generator equipment has a total rating of 10,000 horse-power. From these figures it may be seen that the storage lake would be able to maintain nearly the normal head of water for some hours, when the station was operating under full load, however small the flow of the river above.

FIG. 9.—Canal at Bulls Bridge on Housatonic River.

CHAPTER VII.
THE LOCATION OF ELECTRIC WATER-POWER STATIONS.

Cost of water-power development depends, in large measure, on the location of the electric station that is to be operated. The form of such a station, its cost, and the type of generating apparatus to be employed are also much influenced by the site selected for it. This site may be exactly at, or far removed from, the point where water that is to pass through the wheels is diverted from its natural course.

A unique example of a location of the former kind is to be found near Burlington, Vt., where the electric station is itself a dam, being built entirely across the natural bed of one arm of the Winooski River at a point where an island near its centre divides the stream into two parts. The river at this point has cut its way down through solid rock, leaving perpendicular walls on either side. Up from the ledge that forms the bed of the stream, and into the rocky walls, the power-station, about 110 feet long, is built. The up-stream wall of this station is built after the fashion of a dam, and is reënforced by the down-stream wall, and the water flows directly through the power-station by way of the wheels. A construction of this sort is all that could well be attained in the way of economy, there being neither canal nor long penstocks, and only one wall of a power-house apart from the dam. On the other hand, the location of a station directly across the bed of a river in this way makes it impossible to protect the machinery if the up-stream wall, which acts as the dam, should ever give way. The peculiar natural conditions favorable to the construction just considered are seldom found.

FIG. 10.—Power-house on the Winooski River, near Burlington, Vt.

One of the most common locations for an electric water-power station is at one side of a river, directly in front of one end of the dam and close to the foot of the falls. A location of this kind was adopted for the station at Gregg's Falls, one of the water-powers included in the electric system of Manchester, N. H., where the spray of the fall rises over the roof of the station. Two short steel penstocks, each ten feet in diameter, convey the water from the forebay section of the dam to wheels in the station with a head of fifty-one feet.

FIG. 11.—Canal and Power-house on St. Joseph River, Buchanan, Mich.

A similar location was selected for the station at Great Falls, on the

Presumpscot River (see cuts), whence electrical energy is delivered in Portland, Me. Four steel penstocks, a few feet long and each eight feet in diameter, bring the water in this case from the forebay section of the dam to the wheel cases in the power-house.

FIG. 12.—Power-house on Hudson River at Mechanicsville.

Where the power-station is located at the foot of the dam, as just described, that part which serves as a forebay wall usually carries a head gate for each penstock. The overfall section of a dam may give way in cases like the two just noted without necessarily destroying the power-station, but in times of freshet or very high water the station may be flooded and its operation stopped. The risk of any such flooding will vary greatly on different rivers, and in particular cases may be very slight. Location of the generating station close to the foot of the dam at one end obviously avoids all expense for a canal and cuts the cost of penstocks down to a very low figure.

Such locations for stations are not limited to falls of any particular height, and the short penstocks usually enter the dam nearer its base than its top and pass to the station at only a slight inclination from the horizontal. At Great Falls, above mentioned, the head of water is thirty-seven feet.

A short canal is constructed in some cases from one end of a dam to a little distance down-stream, terminating at a favorable site for the electric station. Construction of this sort was adopted at the Birchem Bend Falls of the Chicopee River, whence energy is supplied to Springfield, Mass. These falls furnish a head of fourteen feet, and the water-wheels are located on the floor of the open canal at its end. The power-station is on the shore side of this canal, and the shafts of the water-wheels extend through bushings in the canal wall, which forms the lower part of one side of the station, to connect with the electric generators inside.

This rather unusual location of water-wheels has at least the obvious advantage that they require no room inside of the station. Furthermore, as the canal is between the station and the river, any break in the canal is not apt to flood the station.

FIG. 13.—York Haven Power-house, on Susquehanna River, Pennsylvania.

An illustration of the use of a very short canal to convey water from one end of a dam to a power-station exists in the 10,000 horse-power plant at Cañon Ferry, Mont., where the head of water is thirty feet. In this case the masonry canal is but little longer than the power-house, and this latter sits squarely between the canal and the river, virtually at the foot of the falls. Other examples of the location of generating stations between short canals and the river may be seen at Concord, N. H., where the head of water is sixteen feet; at Lewiston, Me., where the head is thirty-two feet; and at Spier Falls, on the Hudson River, New York, where there is a head of eighty feet.

There is some gain in security in many cases by locating the power-station several hundred feet from the dam and a little to one side of the main river channel. For such cases a canal may be cheaper than steel penstocks when the items of depreciation and repairs are taken into account. Aside from the question of greater security for the station in the event of a break in the dam, it is necessary in many cases to convey the water a large fraction of a mile, or even a number of miles, from the point where it leaves its natural course to that where the power-station should be located. An example in point exists at Springfield, Mass., where one of the electric water-power stations is located about 1,400 feet down-stream from a fall of thirty-six feet in the Chicopee River, because land close to the falls was all occupied at the time the electric station was built.

FIG. 14.—Power-house at Cañon Ferry on the Missouri River.

FIG. 15.—Shawinigan Falls Power-plant.

The Shawinigan Falls of the St. Maurice River in Canada occur at two points a short distance apart, the fall at one point being about 50 and at the other 100 feet high. A canal 1,000 feet long takes water from the river above the upper of these falls and delivers it near to the electric power-house on the river bank below the lower falls. In this way a head of 125 feet is obtained at the power-house. The canal in this case ends on high ground 130 feet from the power-house, and the water passes down to the wheels through steel penstocks 9 feet in diameter.

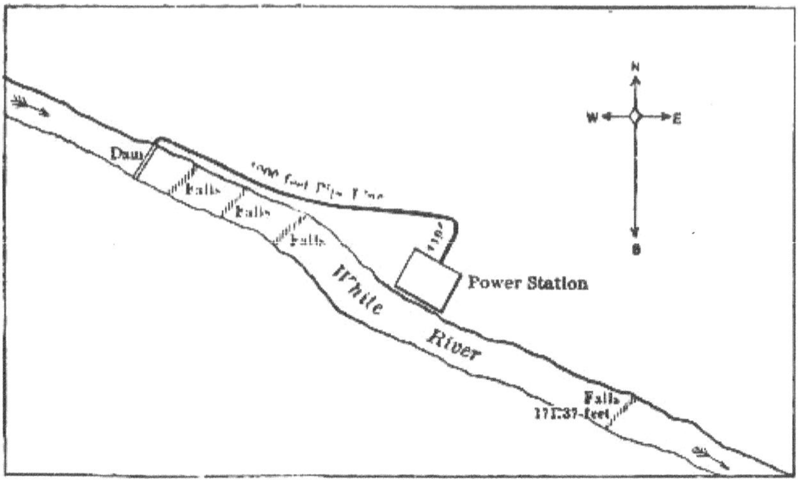

FIG. 16.—Power-house on White River, Oregon.

Another interesting example of conditions that require a power-house to be located some distance from the point where water is diverted from its natural course may be seen at the falls on the Apple River, whence energy is transmitted to St. Paul, Minn. By means of a natural fall of 30 feet, a dam 47 feet high some distance up-stream, and some rapids in the river, it was there possible to obtain a total fall of 82 feet. To utilize this entire fall a timber flume, 1,550 feet in length, was built from the dam to a point near the power-house on the river bank and below the falls and rapids. The flume was connected with the wheels, 82 feet below, by a steel penstock 313 feet long and 12 feet in diameter.

As the St. Mary's River leaves Lake Superior it passes over a series of rapids about half a mile in length, falling twenty feet in that distance. To make the power of this great volume of water available, a canal 13,000 feet long was excavated from the lake to a point on the river bank below the rapids. Between the end of the canal and the river sits the power-station, acting as a dam, and the water passes down through it and the wheels under a head of twenty feet.

FIG. 17.—Power-house Across Canal at Sault Ste. Marie, Mich.

By means of a canal 16,200 feet long from the St. Lawrence River a head of water amounting to fifty feet has been made available at a point on the bank of Grass River near Massena, N. Y. There again the power-station acts as a dam, and the canal water passes down through it to reach the river.

FIG. 18.—Canal and Station on Payette River, Idaho.

From these illustrations it may be seen that in many cases, in comparatively level country, a water-power can be fully developed only by means of canals or pipe lines, and the generating stations cannot be located at the points where the water is diverted.

Thus far the cases considered have been only those with moderate heads and rather large volumes of water. In mountainous country, where rivers are comparatively small and their courses are marked by numerous falls and rapids, it is generally necessary to utilize the fall of a stream through some miles of its length in order to effect a satisfactory development of power. To reach this result, rather long canals, flumes, or pipe lines must be utilized to convey the water to power-stations and deliver it at high pressures.

In cases of this kind the cost of the canal or pipe line may be the largest item in the power development, and it may be an important question whether this cost should be reduced or avoided by the erection of several small generating plants instead of one large one. California offers numerous examples of electric-power development with water that has been carried several miles through artificial channels. An illustration of this class of work exists at the Electra power-house on the bank of the Mokelumne River, in the Sierra Nevada Mountains. Water is supplied to the wheels in this station under a head of 1,450 feet through pipes 3,600 feet long leading to the top of a near-by hill. To reach this hill the water, after its diversion from the Mokelumne River at the dam, flows twenty miles through a canal or ditch and then through 3,000 feet of wooden stave pipe.

Another example of the same sort may be seen in the power-house at Colgate, on the North Yuba River, in the chain of mountains above named. Water taken from this river passes through a wooden flume nearly eight miles long to the side of a hill 700 feet above the power-house, and thence down to the wheels through steel and cast-iron pipes, five in number and thirty inches each in diameter.

Even with long flumes, canals, and pipe lines, it may be necessary to locate a number of generating stations along a single river of the class now under consideration in order to utilize its entire power. Thus on the Kern River, which rises in the Sierra Nevada Mountains and empties into Tulare Lake, two electric power-stations are under construction, and surveys are being made for three more. Of these stations, the one at the lowest level will operate under an 872-foot head of water, and this water, after its diversion from the river, will pass through twenty-one tunnels, with an aggregate length of about ten miles, and through six flumes mounted on trestles and having a total length of 1,703 feet.

Next up-stream is a station near the point where water is diverted for the plant just named. This second station will work under a head of 317 feet, and water for it will come from a point farther up-stream by canals, tunnels, and flumes, with an aggregate length of eleven and one-half miles. At three points still higher up on this river it is the intention to locate three other power-stations by conducting the water in artificial channels, about twelve and one-half, fifteen, and twenty miles in length respectively.

Farther south in California, on the Santa Ana River and Mill Creek, extensive power developments on the lines just indicated have been carried out. On Mill Creek, about six miles from the city of Redlands, is an electric station operating under a head of 530 feet, with water in part diverted from the stream a little less than two miles above and brought down through a steel pipe 10,250 feet long and thirty inches in diameter. This pipe line also takes water from the tail race of another generating plant at its upper end. With some additions and modifications, the station just described is the famous Redlands plant, built in 1893, and believed to be the first for three-phase transmission in the United States.

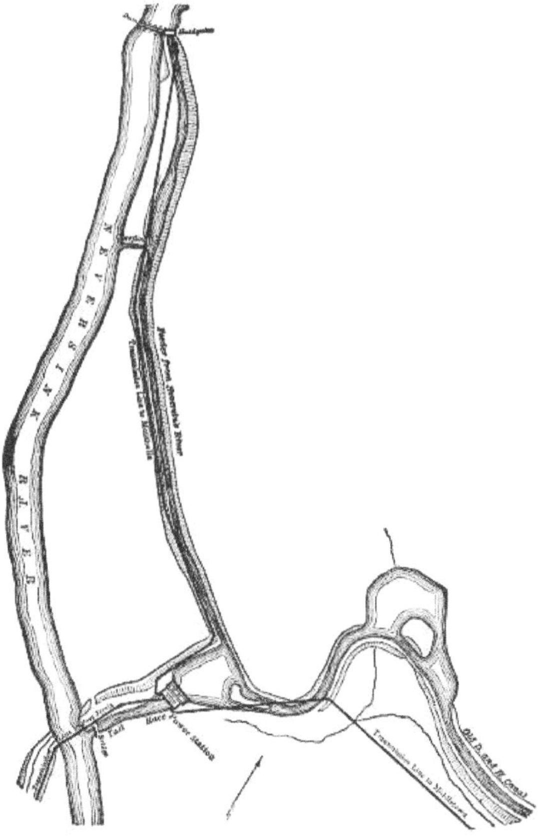

FIG. 19.—Canal and Power-station on Neversink River, New York.

At the upper end of the pipe line just named the second station operates, in part, with water drawn from Mill Creek through a combination of tunnels, flumes, and cement and steel pipes, with a combined length of about three miles, and delivered to some of the wheels with a head of 627 feet. The other wheels at this plant receive water drawn from the same creek by a pipe line about six miles long. A large part of this line is composed of 31-inch cement pipe, laid in trenches and tunnels. The water in the 8,000 feet of pipe next to the power-house has a fall of 1,960 feet, and this pipe is of steel and 24 and 26 inches in diameter. The head of 1,960 feet, minus friction losses in the steel pipes, is delivered at the wheels.

From the foregoing it appears that in a space of eight miles along Mill Creek there is a fall of more than 2,490 feet. To utilize this fall, water is diverted from the creek at three points within a distance of six miles and delivered in two power-stations under three different heads. As the stream gathers in volume between the upper and the lower intakes, an equal amount of power could have been developed in a single station only by taking the three separate conduits or pipe lines to it and delivering their water there at three heads.

Whether the expense of extending conduits and pipe lines to a single generating station will more than offset the advantages to be gained thereby is a question that should be decided on a number of factors varying with each case. In general, it may be said that the smaller the volume of water to be handled and the greater its head, the more advantageous is it to concentrate the generating machinery in the smallest practicable number of stations.

On the Santa Ana River, into which Mill Creek flows, the Santa Ana plant, whence energy is transmitted to Los Angeles, is located. Water reaches this plant through a conduit of tunnels, flumes, and pipes, with a total length of about three miles from the point where the flow of the river is diverted. The 2,210 feet of this conduit nearest the power-plant are composed of 30-inch steel pipe, with a fall of 728 feet.

Within fifteen miles of Mexico City are five water-power stations that supply energy for its electrical system. Two of these stations are on the Monte Alto and three are on the Tlalnepantla River, the two former stations being about three miles, and the more distant of the three latter stations five miles, apart. At a distance of several miles above the highest station on each river the water is diverted by a canal, and the water of each of these canals, after passing through the wheels of the highest station, goes on to the remaining station, or stations, on the same river by a continuation of the canal.

FIG. 20.—Wood Pipe Line to Pike's Peak Power-house.

By placing the stations so short a distance apart the head of water at each station is reduced. On one stream these heads are 492 and 594 feet respectively, and at two of the stations on the other stream they are 547 and 295 feet respectively. This division of the total head of water afforded by each river results in a rather small capacity for each station, the total at the five plants being only 4,225 kilowatts.

In contrast with this figure the already mentioned Electra plant has generators of 10,000, the Santa Ana plant generators of 3,000, and the larger of the two Mill Creek plants generators of 3,500 kilowatts capacity. It should be noted that the cost of operation, as well as that of original construction, will vary materially between one large and several smaller stations of equal total capacity, the advantage as to operative cost being obviously with the one large plant.

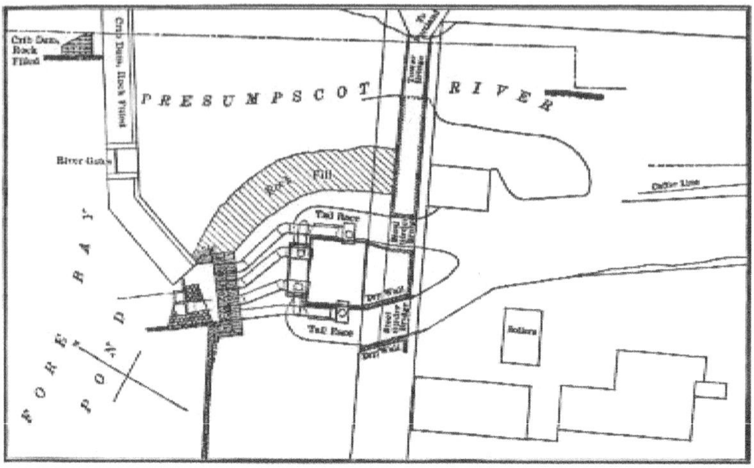

FIG. 21.—Power-house at Great Falls, Presumpscot River.

All of the power-stations here considered have been equipped with water-wheels and generators operating on horizontal shafts, and this is the general practice. This arrangement brings the generators and the floor of the power-station within a few feet of the level of the tail-water. By the general use of draught tubes with turbine wheels the floors of stations are often kept twenty feet or more above the tail-water level.

Where the total available head of water is quite small, as is often the case with rivers where the volume of water is great, it is generally necessary to bring the level of the station floor down to within a few feet of the tail-water. The Birchem Bend station of the Springfield, Mass., electric system affords a good example of this sort, the floor of this station being only 2.6 feet above the ordinary level of the tail-water. At this station the difference of level between the head- and tail-water is only fourteen feet, and even with the low floor level named the top sides of the horizontal turbine wheels are covered only by 4.5 feet of water.

At the Garvin's Falls station of the Manchester, N. H., electric system the level of the floor of the generator room is thirteen feet above the ordinary level of the Merrimac River, on the bank of which this station is located; but in this case the total head of water is about twenty-eight feet. The high water of the Merrimac in 1896, before the Garvin's Falls station was built, reached a point 5.24 feet above its present floor level, and 18.24 feet above the ordinary level of the river at the point where the station is located.

Under the Red Bridge electric station of the Ludlow Manufacturing Company, on the Chicopee River, in Massachusetts, the tail-water is twenty feet below the level of the floor and twenty-four feet below the centres of the water-wheel and generator shafts. The difference between wheel-shaft and tail-water levels at this station is near the maximum that can be attained with horizontal pressure turbines, because a draught tube much longer than twenty-five feet does not give good results.

In a pressure turbine the guides and wheel must be completely filled with water, as must also the draught tube, for efficient operation. If draught tubes are much more than twenty-five feet long, it is hard to keep a solid column of water from turbine to tail-water in each, and if this is not done a part of the head of water becomes ineffective. As pressure turbines are employed almost exclusively at electric stations with low heads of water, it is frequently impossible to locate such stations above the possible level of tail-water in times of flood if horizontal wheels direct-connected to generators are employed.

FIG. 22.—Power-house at Garvin's Falls on the Merrimac River.

If turbines with vertical shafts are to be used, a power-station may be so located or constructed that all the electrical equipment will be above the highest known water-mark. With vertical shafts, connecting wheels, and generators, the main floor of an electric station may be located above the crest of the falls where the power is developed instead of at or near their base.

FIG. 23.—Power-house No. 2 at Niagara Falls.

By far the most important examples of electric stations laid out on this plan are those at Niagara Falls, where there are four such plants. Two of these generating plants, with an aggregate capacity of 105,000 horse-power, stand a mile above the falls, and are supplied with water through a short canal from Niagara River. Beneath each of these two stations a long, narrow wheel pit has been excavated through rock to a depth of 172 feet below the level of water in the canal. Both wheel pits terminate in a tunnel 7,000 feet long that opens into the river below the falls.

In this wheel pit the tail-water level is 161 feet below that of the water in the canal, and 166 feet below the floor of the power-station. Water passes from the canal down the wheel pits to the wheels near the bottom through steel penstocks, each seven feet in diameter, and a vertical shaft extends from each wheel case to a generator in the station above.

Locations like that at Niagara give great security against high water and washouts, but are seldom adopted because of the large first cost of plant construction. With heads of water from several hundred to 2,000 feet the loss of a few feet of head reduces the available power to only a very slight extent, and impulse wheels are usually employed. Draught tubes are not available to increase the heads at such wheels, and any fall of the water after it leaves the wheels does no useful work.

FIG. 24.—Colgate Power-house.

Electric stations driven by impulse wheels under great heads, like those at Colgate, Electra, Kern River, Santa Ana River, and Mill Creek, may be located far enough above the beds of their water-courses to avoid dangers from freshets, without serious loss of available power.

CHAPTER VIII.
DESIGN OF ELECTRIC WATER-POWER STATIONS.

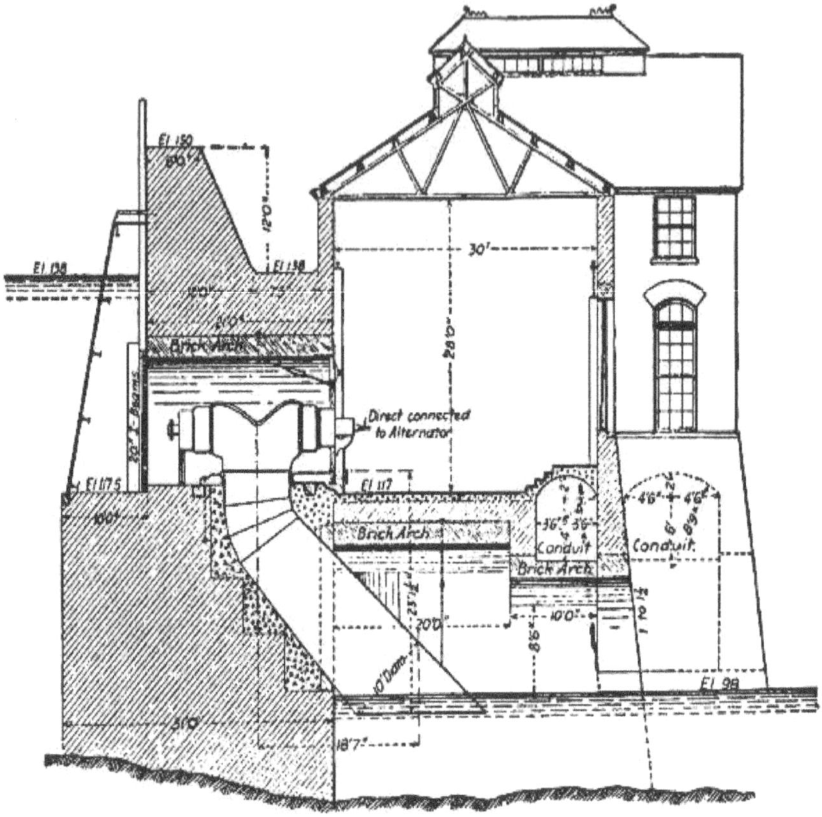

FIG. 25.—Cross Section of Columbus, Ga., Power-station.

Water-wheels must be located at some elevation between that of head- and tail-water. With horizontal shafts and direct-connected wheels and generators the main floor of the station is brought below the level of the wheel centres. This is much the most general type of construction, and was followed in the Massena, Sault Ste. Marie, Cañon Ferry, Colgate, Electra, Santa Ana, and many other well-known water-power stations. If horizontal shafts are employed for wheels and generators with belt or rope connections between them the floor of the generator room may be elevated a number of feet above the wheels. This difference of elevation is usually provided for either by upper and lower parts of the same room, or by separate rooms one above the other and a floor between them. A two-story construction of this latter sort was frequently adopted in the older water-power stations, and good examples of it may be seen in connection with the electrical supply system at Burlington, Vt., and the Indian Orchard station in the Springfield, Mass., system. Vertical wheel shafts make the elevation of the main or generator floor of a station independent of that of the wheels, and thus give the highest degree of security against high water. After the vertical wheel shaft reaches the generator room, it may be geared to a horizontal shaft that has one or more dynamos directly mounted on it, or drives dynamos through belts or ropes. Belt-driving in this way, from horizontal shafts connected by bevel gears with vertical wheel shafts, is not uncommon in the older class of water-power stations. Generators mounted singly or in pairs on horizontal shafts that are driven by gearing on vertical wheel shafts have been adopted at the Lachine Rapids and South Bend plants, and it seems to offer a desirable method of connection in cases where vertical wheels are necessary and the cost of generators must be kept at a low figure. With this method of driving the generators can be designed for any economical speed and step bearings avoided.

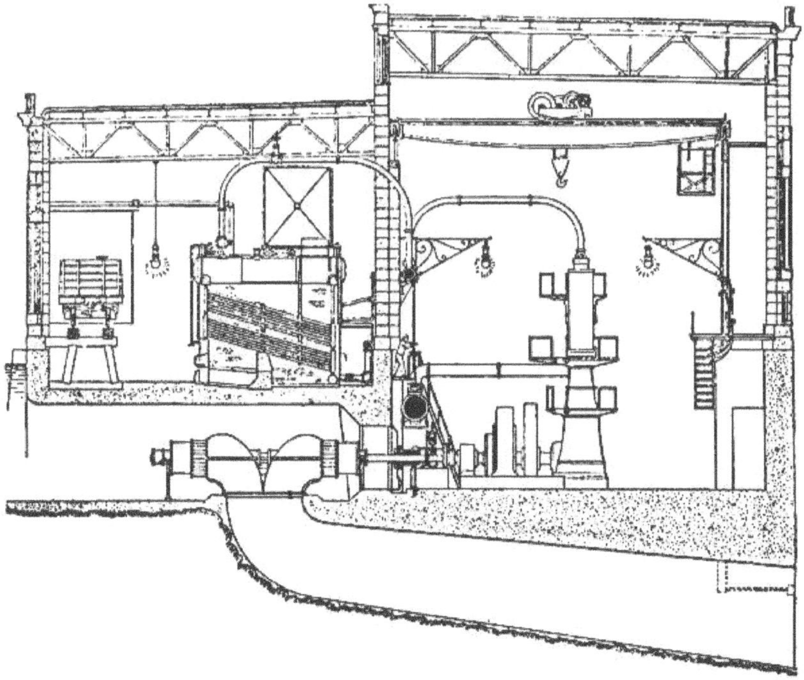

FIG. 26.—Cross Section of Combined Steam- and Water-power Station at Richmond, Va.

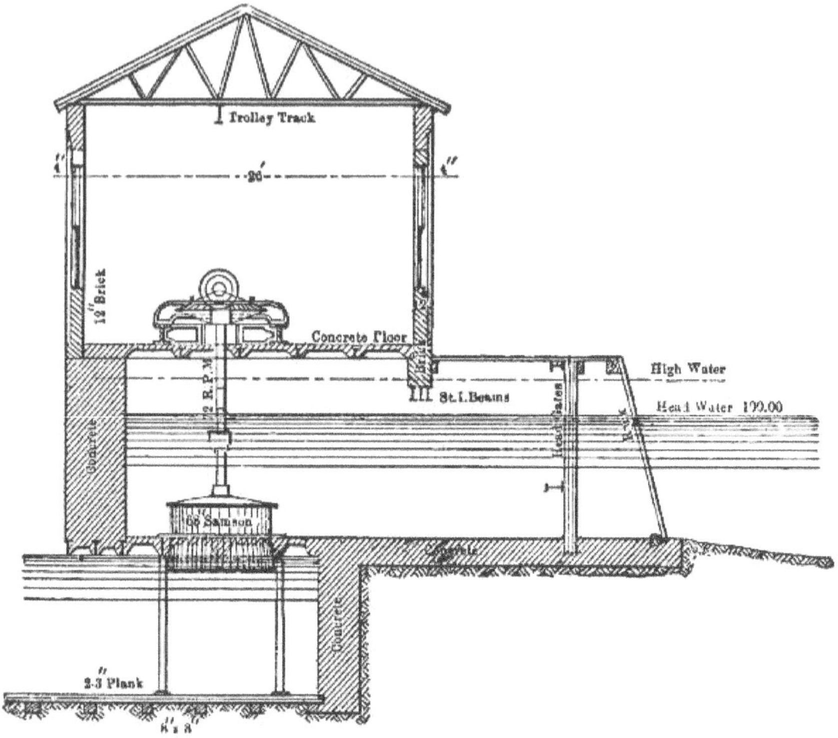

FIG. 27.—Cross Section of Wheel House at Buchanan, Mich.

The most desirable method of driving generators with vertical wheels, where the expense is not too great, is the direct mounting of each generator on the upper end of a wheel shaft (see cut). This method of connection not only requires a special type of generator, but may put serious limits on its speed. In general, the peripheral speed of a pressure turbine should be about 75 per cent of the theoretical velocity of water issuing under a head equal to that at which the wheel operates, in order to give the best efficiency. The rotative speeds of turbines, operating under any given head, should thus increase as their capacities and diameters decrease. Because of these principles it is the common practice, with horizontal wheels, to mount two or more on each shaft to which a generator is direct-connected in order to obtain a greater speed of rotation than could be obtained with a single wheel of their combined power. Thus, at Sault Ste. Marie the horizontal shaft on which each 400-kilowatt generator is mounted is driven at 180 revolutions per minute by four turbines under a head of about 20 feet. At Massena the head of water is 50 feet, and each 5,000 horse-power generator is driven at 150 revolutions per minute by six turbines on a horizontal shaft. Vertical turbines are sometimes mounted singly on their shafts, as was done in the hydroelectric plant at Oregon City on the Willamette River, and this practice gives speeds that are too low for direct-connected dynamos of moderate cost, unless the head of water is unusually great. At the Oregon City plant the head of water is only 40 feet, and yet a single 42-inch turbine was mounted on the vertical shaft that drives each generator.

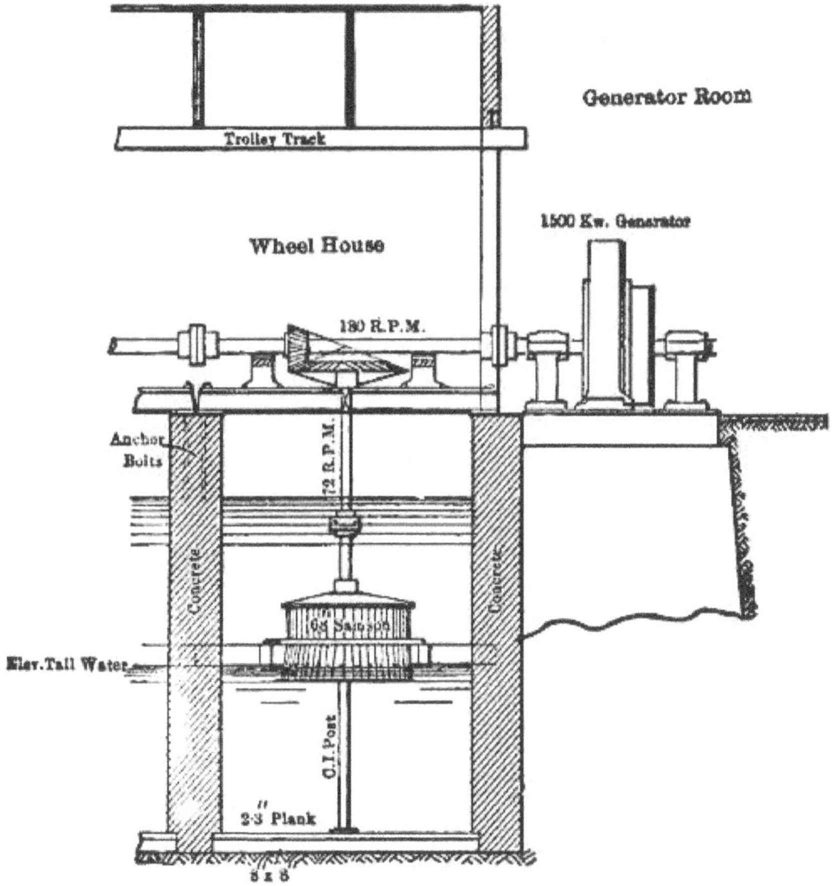

FIG. 28.—Longitudinal Section of Buchanan, Mich., Power-house.

The most notable examples of direct-connected generators and vertical turbines is that at Niagara Falls, where twenty-one generators of 5,000 horse-power each are mounted at the tops of as many vertical wheel shafts in two of the four stations. Each vertical shaft in the Niagara stations is driven at 250 revolutions per minute by a pair of turbines, one above the other. The maximum head between the water in the Niagara canal and that in the tunnel which forms the tail-race is 161 feet. On ten shafts the centres of the wheel cases are 136 feet below the level of water in the canal, and no draft tubes are used.

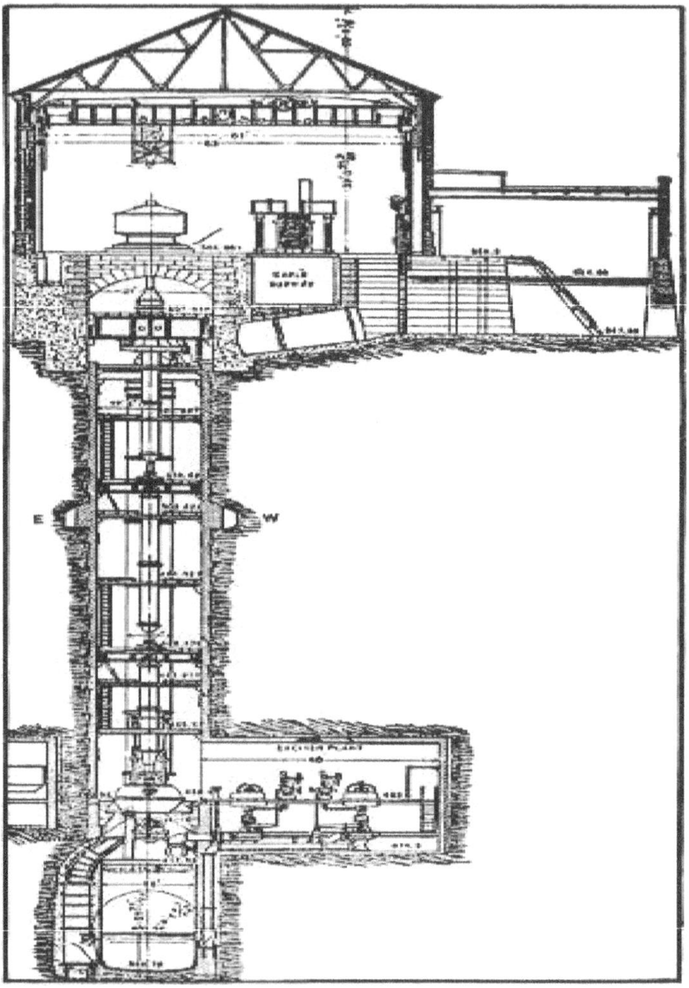

FIG. 29.—Section of Power-house No. 2 at Niagara Falls.

The eleven pairs of wheels at the second Niagara power-house have their centre line 128.25 feet below the canal level and a draft tube for each pair of wheels extends to a point below the tail-water level. It is entirely practicable to use more than a single pair of turbines on the same vertical shaft, as is shown at the Hagneck station on the Jura, in Switzerland, where the head of water is about twenty-one feet and four turbines are mounted on each vertical shaft. The combined capacity of these four wheels on each shaft is

1,500 horse-power and its speed is 100 revolutions per minute. At the top of each shaft an 8,000-volt generator, with external, revolving magnet frame, is mounted. The use of four wheels per vertical shaft presents no great difficulty and should be resorted to more frequently in the future.

FIG. 30.—Interior of Power-house, Buchanan, Mich.

For horizontal, direct-connected turbine wheels and generators the nearly uniform practice is to locate the generators in a single row from one end of a station to the other, and this brings the turbines into a parallel row. On this plan the shaft of each connected generator and its group of turbines sets at right angles to the longer sides of a station and approximately parallel with the direction in which water flows to the wheels. The typical water-power station with direct-connected units is thus a rather long, narrow building into which water enters on one side through penstocks and leaves on the other through tail-races. Such stations usually set with one of the longer sides parallel to the river into which the tail-water passes and between this river and the canal or pipe line. At Massena the electric station occupies the position of a dam between the end of the power canal and the Grass River, being about 150 feet wide and 550 feet long. Canal water entering this station passes through its wheels to the river under a head of about 50 feet. A similar construction was followed at Sault Ste. Marie, where the power-station separates the end of the canal from the St. Mary's River. This station is 100 feet wide, 1,368 feet long, and is to contain 80 sets of horizontal wheels, each set being connected to its own generator, and through these wheels the canal water passes under a head of approximately 20 feet. Ten generators are placed in line at the Cañon Ferry station which is 225 by 50 feet inside, and each generator is driven by a pair of horizontal wheels under a head of 30 feet. This station sets between a short canal and the Missouri River, near one end of the dam. Passing from water-heads of less than 50 to those of several hundred or even more than 1,000 feet, the general type of station building remains about the same, but there is an important change in the arrangement of direct-connected wheels and generators. With these high heads of water, wheels of the impulse type, to which the water is supplied in the form of jets from nozzles, are employed. These jets pass to the wheels in planes at right angles to their shafts, instead of flowing in lines parallel to these shafts like water to pressure turbines. The shafts of impulse wheels and their direct-connected generators are consequently arranged parallel with the longer instead of the shorter sides of their stations. This plan results in long, narrow stations with water entering at one and leaving at the other of the longer sides, just as in the case of direct-connected turbines under moderate heads. Stations with direct-connected impulse wheels are even longer for a given number and capacity of units than are stations with pressure turbines. Colgate power-house, on the North Yuba River, contains seven generators, each direct-connected to an impulse wheel and shafts all parallel to its longer sides. This station is 275 feet long by 40 feet wide, and the water which enters one side by five iron pipes, 30 incheseach in diameter, under a head of about 700 feet, is discharged from the other side into the river.

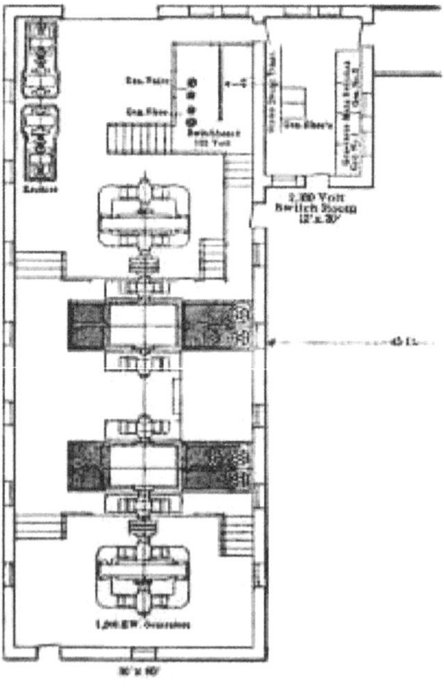

FIG. 31.—Plan of Generating Station near Cedar Lake for City of Seattle, Wash.

FIG. 32.—Foundation of Power-station at Spier Falls.

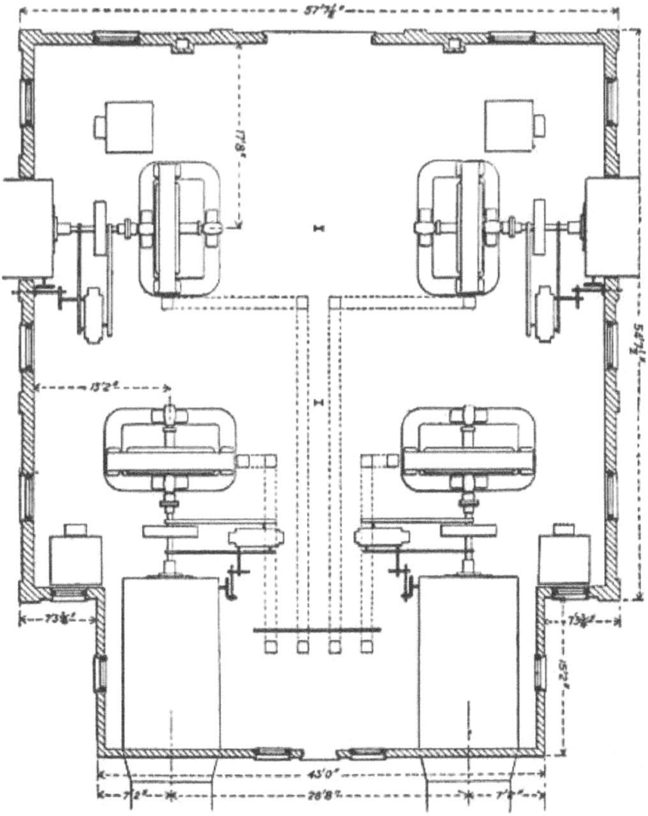

FIG. 33.—Plan of Power-station at Great Falls.

At Electra station on the Mokelumne River five pairs of impulse wheels are direct-connected to five generators, each unit having its shaft diagonal with the walls of the building, and pipes deliver water to the wheels under a head of 1,450 feet. The ground plan of the generator room at this plant is 40 by 208 feet. The power-station on Santa Ana River, whence energy is transmitted 83 miles to Los Angeles, measures 127 feet long and 36 feet wide inside, and contains four generating units in line, each of which consists of a direct-connected dynamo and impulse wheel, with shafts parallel to the longer sides of the station. Jets driving the wheels in this station are delivered under a head of 728 feet minus the loss by friction in a penstock 2,210 feet long.

FIG. 34.—Power-house at Red Bridge on Chicopee River.

Both of the first Niagara plants, with vertical wheels far below the stations in the pits, are long and narrow and have their generators in a single row. The later of these two stations has a ground area of approximately 72 by 496 feet outside, and contains eleven generators all in line. From these examples it may be seen that the prevailing type of electric water-power station, whether designed for horizontal or vertical wheels of either the pressure or impulse type, is wide enough for only a single row of generators and wheels, and has sufficient length to accommodate the required number of units.

A few modern stations that depart from this general plan will be found, as that at Great Falls, on the Presumpscot River, whence electrical supply for Portland, Me., is drawn. This station sets about forty feet in front of the forebay end of the dam, and two penstocks enter the rear wall, while the other two enter one each through two of the remaining opposite sides. Of the four generators, with their direct-connected wheels, two are arranged with parallel shafts, while the other two have their shafts in line and at right angles to the lines of the former two. The station containing these generating sets has a floor area of 55 by 67.5 feet.

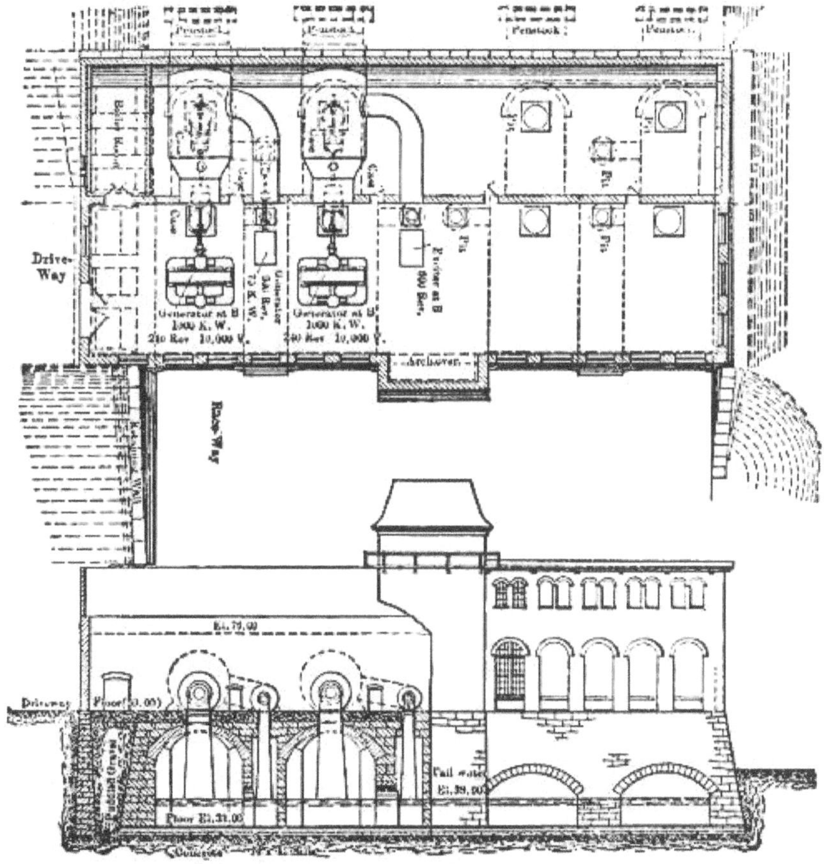

FIG. 35.—Plan and Elevation of Red Bridge Station on the Chicopee River.

Modern electric stations driven by water-power are usually but one story in height and are clear inside from floor to roof, save for cranes and roof trusses. This construction may be seen in the Niagara, Spier Falls, Cañon

Ferry, Colgate, Electra, Santa Ana River and many other notable plants. In spite of this one-story style of construction, the electric stations reach fair elevations because of the necessity for head room to operate cranes in placing and removing generators. At Garvin's Falls, on the Merrimac River, the electric station contains generators of 650 kilowatts each and the distance from floor to the lower cords of roof trusses is 27 feet. In the station at Red Bridge, on the Chicopee River, where generators are of 1,000 kilowatts capacity each, the distance between floor and the under side of roof beams is 30.66 feet. Between the floor and roof trusses at the Birchem Bend station, on the river last named, the distance is 26.25 feet, but each generator is rated at only 400 kilowatts. In the Cañon Ferry plant, with its generators of 750 kilowatts each, the distance from floor to roof trusses is 28 feet. At the plant on Santa Ana River, the 750-kilowatt generators, being connected to impulse-wheels, operate at 300 revolutions per minute, have relatively small diameters and are mounted over pits in the floor so that their shaft centres are only about two feet above it. By these means the distance from floor to roof trusses was reduced to 18.25 feet. All these examples of elevations between floors and roof supports are for stations with direct-connected generators and horizontal wheels. In the new Niagara station, where generators of 3,750 kilowatts each are mounted on vertical wheel shafts that rise from the floor, the distance between the floor and roof trusses is 39.5 feet.

Electric stations driven by water-power are now constructed almost entirely of materials that will not burn—that is, stone, brick, tile, concrete, cement, iron, and steel. Stone masonry laid with cement mortar or concrete masonry is very generally employed for all those parts of the foundations that come in contact with the tail-water. For sub-foundations bedrock is very desirable, but where this cannot be reached piles are driven closely and their tops covered with several feet of cement concrete as a bedding for the stone foundation. Where stone is plenty or bricks hard to obtain, the entire walls of a water-power station are frequently laid entirely with stone in concrete mortar. If bricks can readily be had they are more commonly used than stone for station walls above the foundations. Concrete formed into a monolithic mass is a favorite type of construction for the foundations, walls and floors of water-power plants in Southern California. Cement and concrete are much used for station floors in all parts of the country, and these floors are supported by masonry arches in cases where the tail-water flows underneath the station after leaving the wheels. Station roofs are usually supported by steel trusses or I-beams, and slate and iron are favorite roof materials. With iron roof-plates an interior lining of wood, asbestos, or some other poor conductor of heat is much used to prevent the condensation of water on the under side of the roof in cold weather. Walls of water-power stations are usually given sufficient thickness of masonry to support all loads that come upon them without the aid of steel columns. In some cases where cranes do not extend entirely across their stations, one end of each crane is supported by one of the station walls and the other end by a row of iron or steel columns rising from the floor. Where the generator-room of a station has its floor level below high-water mark especial care should be taken to make the walls water-proof to an elevation above this mark. As the travelling-crane and the loads which it carries in erecting wheels and generators form a large part of the weight on the station walls, these walls are often reduced as much as one-half in thickness at the level of the crane, thus forming benches on which the ends of the cranes rest.

FIG. 36.—Steel Penstocks at Chamblay Power-house.

The Garvin's Falls station, on the Merrimac River, rests on arches of stone masonry through which the tail-water passes, and the brick walls are water-proofed to an elevation eight feet above the floor. At twenty feet above the floor the twenty-four-inch brick walls on the two longer sides are reduced to eight inches in thickness, thus forming benches each sixteen inches wide on which the crane travels. Arches of stone masonry support the twenty-four-inch brick walls of the station at Red Bridge, on the Chicopee River, and these walls on the two longer sides decrease in thickness to twelve inches at an elevation of twenty-one feet above the floor, thus forming benches twelve inches wide for the ends of the crane.

One concrete wall of the Santa Ana station is 2.5 feet thick to a distance of 13.5 feet above the floor, and then shrinks to a thickness of 1.5 feet, corresponding to that of the opposite wall, thus forming a bench twelve inches wide for one end of the crane. The other end of the crane in this case is supported by an I-beam on a row of iron columns.

It is not uncommon to locate horizontal turbines in a room separate from that occupied by the generators to which they are direct-connected, in order to protect the latter from water in the event of a break in penstocks or wheel cases. In cases of this sort the shafts connecting wheels and generators pass through the wall between them. The horizontal turbines may be located at the bottom of a canal whose water presses against the wall through which the wheel shafts pass, or they may be contained in iron cases at the ends of penstocks. In this latter case an extension of the station is often provided for a wheel room to contain these cases. Such wheel rooms are long, narrow, low-roofed and parallel to the generator rooms of their stations. The floors of these wheel rooms are at nearly the same levels as the floors of generator rooms, but elevations of their roofs above the floors are much less than like elevations in the main parts of the stations. The Garvin's Falls, Red Bridge, and Apple River stations have wheel rooms of the type just described. With impulse-wheels to which water passes in planes at right angles to their shafts it is desirable, in order to avoid changes in the direction of water pipes, that direct-connected wheels and generators occupy the same room, and this is the arrangement at the Colgate, Electra, Santa Ana, Mill Creek, and many other power-houses using such equipments. The area of a wheel room may frequently be reduced at stations operating direct-connected horizontal-pressure turbines under low heads by placing the wheels at the bottom of the canal which has one side of the station or generator room for a retaining wall. This plan was adopted at the Birchem Bend plant with a head of fourteen feet, and at the Sault Ste. Marie station where the head of water is about twenty feet. Vertical wheels direct-connected to generators must be directly underneath the main room of their station, and may be in a canal over which the station is built, in a wheel room that forms its lower part, or in a wheel pit and supplied with water through penstocks, as at the Niagara Falls plants.

Step-up transformers developing very high voltages are not an element of safety in a generator room, and the better practice is to locate them in a separate apartment by themselves, if not in a separate building. For the Niagara Falls plant, the transformers that deliver three-phase current at 22,000 volts are located in a building across the canal from the generating plant. At Cañon Ferry the transformers operating at 50,000 volts, three-phase, are located in a steel and iron addition to the power-house.

Transformers at Electra station, which are intended to work ultimately at 60,000 volts, are located in an extension of the main building and are separated from the generator-room by a wall. At the Santa Ana plant the 33,000-volt transformers are grouped in one corner of the generator room, but no partition separates their space from the remainder of the room. In the Colgate plant the transformers, working at 40,000 volts, are spaced along one of the longer sides of the station opposite to and only a few feet from the row of generators. One end of the main room in the Apple River plant is devoted exclusively to the 25,000-volt transformers, and there is a distance of about twenty-seven feet between them and the nearest generator. The highest degree of safety for transformers at these great voltages seems to require that they be located in a separate room where the floor, walls, and roof are made entirely of incombustible material.

FIG. 37.—One of the Turbine Wheels at Spier Falls on the Hudson River.

Water supplied to horizontal turbine wheels under moderate heads usually enters the station by penstocks on one side and leaves it by the tail-race on the other, but this is not true in every case. At the Birchem Bend plant, the canal in which the wheels are located being between the station and the river, water never enters or passes under the station, which has a continuous foundation. So again at the Apple River plant the single supply pipe, twelve feet in diameter and delivering water under a head of eighty-two feet, lies parallel with the greater length of the station and between it and the river. Short penstocks pass from this supply pipe into the wheel section of the power-house, and the water after passing through the wheels flows out to the river between the masonry piers that support the twelve-foot pipe. The generator section of this station has thus no water flowing under it. An interesting distinction may be noted between the conditions as to the tail-water about the foundations of stations working under low and those under great water heads. In cases of the former sort the volumes of water are relatively great and the foundations of stations are usually submerged, and much reduced in area to make room for the tail-races. Thus, the foundations of the station at Red Bridge, where there is 49 feet head, have nearly all of their footings under water, and of a total length of 145 feet at the top of these foundations the six tail-races underneath cut out 92 feet. These tail-races extend underneath both the wheel and generator rooms.

Where power is derived from water delivered under great head from pipe nozzles to impulse-wheels, stations are usually well above the water levels of streams into which they discharge, and passages for tail-water underneath the station shrink to small tunnels through their foundations. Seven of these tunnels have a total width of less than 25 feet at the Santa Ana River station, which is 127 feet long, and where the head of water is 728 feet. At the Colgate plant, with its head of 700 feet, the water, at times of light load, instead of flowing out of its passages underneath the station, shoots from the pipe nozzles clear across the North Yuba River on the bank of which the station stands.

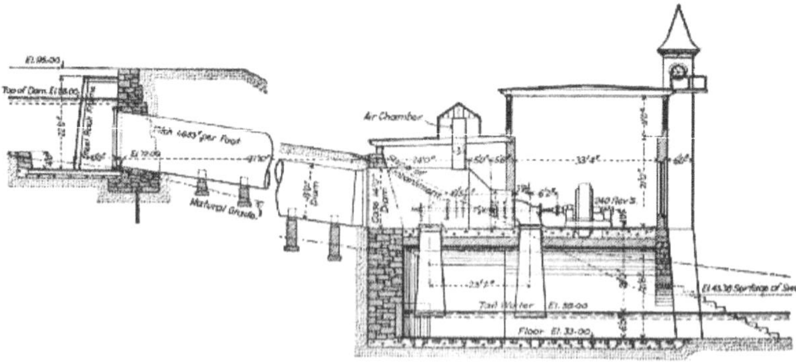

FIG. 38.—Power Plant of Ludlow Manufacturing Company.

In a comparison of floor areas per kilowatt of main generator capacities in electric stations using water- and those using steam-power, the matter of space for transformers may be entirely omitted, because the extent of this space is independent of the type or location of water-wheels, or the difference of water and steam as motive powers. Where water-wheels and their connected generators occupy separate rooms, as is often the case with turbines under low pressures, the wheel room has a little less length, and is generally narrower than the generator room. Thus, at the Red Bridge station the generator room is 141 feet long and the wheel room about 127 feet, while the former is 33.33 feet and the latter 24 feet wide. So again at Apple River Falls the generator room is 140 by 30 feet and the wheel room 106 by 22 feet, the generator room in this case containing also transformers. It follows that if wheels can be located outside of the station, as in a canal, quite a reduction in its total floor area can be made, which may easily range from 20 to 40 per cent. The kilowatt capacity per square foot of floor area in both wheel and generator rooms combined tends to increase with the individual capacity of the generating units. Generators on vertical shafts seem to require about as much floor space per unit of capacity as do generators on horizontal shafts. In the Red Bridge station the total capacity is 4,800 kilowatts of main generators in six horizontal units, and the area of the generator room alone is 0.96 square foot per kilowatt of this capacity. The second station with vertical units at Niagara Falls has a capacity of 41,250 kilowatts in eleven generators on vertical shafts, and its floor area amounts to 0.86 square foot per kilowatt; narrow impulse-wheels of large diameter tend to economy of floor space, as in Electra station, where the room containing wheels and generators has an area of only 0.83 square foot per unit of its 10,000 kilowatts capacity. At the Colgate plant, where the total rating of generators is 11,250 kilowatts, the floor area under wheels and generators is almost exactly one square foot per kilowatt. The Santa Ana station, with a total capacity of 3,000 kilowatts, has 1.52 square feet of floor area for each unit of capacity. This last figure may be compared with the 1.72 square feet per kilowatt of generator rating for the 4,800-kilowatt station at Red Bridge and the 1.75 square feet per unit of capacity in the 800-kilowatt plant at Birchem Bend.

FIG. 39.—Power-house on Payette River, Idaho.

All types of water-power stations with direct-connected wheels and generators have much smaller floor areas per unit capacity than do steam-power stations with direct-connected horizontal units. Thus, the modern steam-driven station at Portsmouth, N. H., has a plan area in engine- and boiler-rooms of 16,871 square feet, and its total capacity in four direct-connected units is 4,400 kilowatts, so that the area amounts to 3.82 square feet per kilowatt rating of its generators. Of this area about 46 per cent is in the boiler-room.

FLOOR DIMENSIONS FOR DIRECT-CONNECTED, HORIZONTAL WATER-WHEELS AND GENERATORS AT ELECTRIC STATIONS.

Station.	Feet Long.	Feet Wide.	Number of Generators.	Total Kilowatt Capacity.
[A]Niagara, No. 2	496	72	11	41,250
Sault Ste. Marie	1,368	100	80	32,000
Colgate	275	40	7	11,250
Electra	208	40	5	10,000
Cañon Ferry	225	50	10	7,500
Red Bridge	141	57	6	4,800
Apple River	140 106	30 22	4	3,000
Santa Ana River	127	36	4	3,000
Great Falls	67.5	55	4	2,000
Garvin's Falls	62 50	30 23	2	1,300
Birchem Bend	56.6	26.7	2	800
Portsmouth (steam-driven)	14.4 inside, but minus 360 square feet.	119.66	5	4,400

[A] Vertical wheel shafts.

Some of these dimensions apply to the inside and some to the outside of stations. Some small projections are not included.

CHAPTER IX.
ALTERNATORS FOR ELECTRICAL TRANSMISSION.

Dynamos in the generating station of an electric transmission system should be so numerous that if one of them is disabled the others can carry the maximum load. If only two generators are installed, it is thus desirable that each be large enough to supply the entire output, so that the dynamo capacity exceeds the greatest demand on the station by 100 per cent. To avoid so great excess of dynamo capacity it is common practice to install more than two generators.

Other considerations also tend to increase the number of dynamos in the generating station of a transmission system. Thus one transmission line may be devoted exclusively to lighting, another to stationary motors, and a third to electric railway service; and it may be desirable that each line be supplied by an independent dynamo to avoid any effect of fluctuations of railway or motor load on the lighting system.

At the generating station of the transmission system that supplies electric light and power in Portland, Me., the idea of independent units has been carried out with four 500-kilowatt dynamos, each driven by a pair of wheels fed with water through a separate penstock from the dam. Each of these dynamos operates one of the four independent transmission circuits. Where a number of water-power stations feed into a single sub-station the requirement that each generating station have its capacity divided up among quite a number of dynamos may not exist, since one station may be entirely shut down for repairs and the load carried meantime by the other stations. A good illustration of this point may be seen at Manchester, where a single sub-station receives energy transmitted from four water-power plants. At one of these plants the entire capacity of 1,200 kilowatts is in a single generator.

The foregoing considerations as to the number of dynamos apply with equal force to both steam- and water-driven stations, but other factors tend to increase the number of dynamos in water-power plants where the head of water is comparatively small. This tendency is due to the fact that the peripheral speeds of pressure turbine water-wheels should be about twenty-five per cent less than the velocity at which water would issue from an opening under the head of water at which these wheels operate in order to secure high efficiency. This velocity of water and therefore the peripheral speed of pressure turbine wheels varies with the square root of the head of water.

FIG. 40.—Generators at Sault Ste. Marie Power Plant.

Since the peripheral speed of turbines is thus determined by the heads of water under which they operate, and since the diameters of turbines must increase with their capacities, the rate of revolution for pressure turbines under any given head decreases as the power goes up. For this reason it is often desirable to use a larger number of dynamos in a water-power plant than would otherwise be required in order to avoid very low speeds of revolution on the direct-connection to the turbines. A notable illustration of this practice exists in the great water-power plant of the Michigan-Lake Superior Power Company, at Sault Ste. Marie, Mich., where a generating capacity of 32,000 kilowatts is divided up between 80 dynamos of 400 kilowatts each. The head of water available at the pressure turbines in this plant is about 16 feet, and their speed is 180 revolutions per minute. In order to obtain even this moderate speed under the head of 16 feet it was necessary to select turbines of only 140 horse-power each. Four of these turbines are mounted on each shaft that drives a 400-kilowatt dynamo, direct-connected, so that there are 320 wheels in all. Had a smaller number of wheels been employed to yield the total power their speed and that of direct-connected dynamos must have been less than 180 revolutions per minute. As the cost of dynamos increases with very low speeds it is often cheaper to install a larger number of dynamos at a higher speed than a smaller number at a lower speed for a given total capacity.

The use of a larger number of units than would otherwise be necessary in order to avoid a very low speed is further illustrated by the 7,500-kilowatt plant of the Missouri River Power Company, at Cañon Ferry, Mont. This capacity is made up of ten generators, each rated at 750 kilowatts and direct-connected to a pair of pressure turbine wheels operating at 157 revolutions per minute, under a head of about 32 feet.

Under comparatively high heads of water pressure turbines operate at speeds that are ample for direct-connection to even the largest dynamos.

FIG. 41.—Interior of Power-house No. 2, Niagara Falls.

Thus in the Niagara Falls plant, where the head of water is 136 feet, each pair of turbines drives a direct-connected dynamo of 3,750 kilowatts at 250 revolutions per minute. In the rare case where the power to be developed is so great that the number of generators necessary to give security and reliability to the service leaves each generator with a capacity larger than is desirable for structural reasons, the number must be increased simply to reduce the size of each generator. Such a state of facts existed at Niagara Falls, where the first station contains ten dynamos of 3,750 kilowatts each, and the second station contains eleven units of like capacity.

In the greater number of transmission systems the generators are direct-connected to either steam-engines or water-wheels, and their speeds of rotation are largely determined by the requirements of these prime movers. Steam-engines can be designed with some regard to the desirable speeds for direct-connection to dynamos, but water-wheels are less flexible in this particular. Each type of wheel has its peripheral speed mainly determined by the head of water under which it may be required to operate, and variation from this speed means serious loss of efficiency.

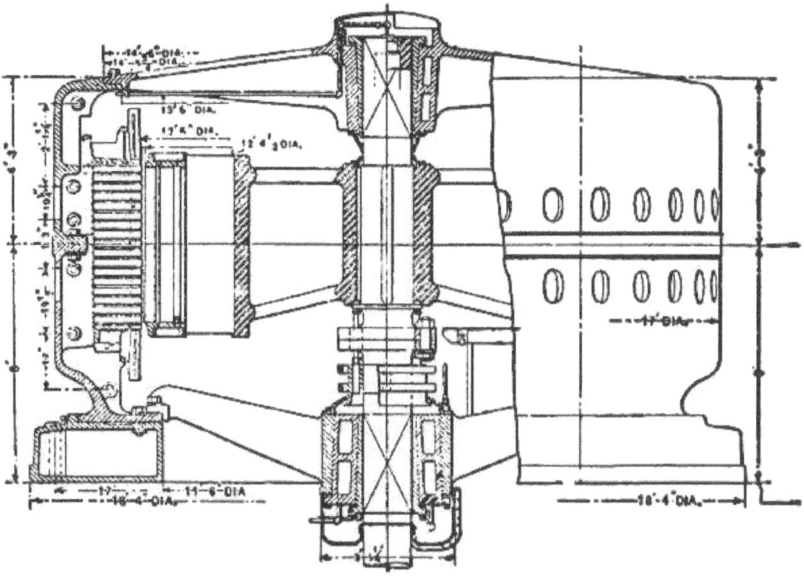

FIG. 42.—10,000 H. P. 12,000 Volt Generator in Canadian Power-house at Niagara Falls.

Under heads of much more than 100 feet pressure turbines operate at rather high speeds in all except very large sizes. It is much the more common to see water-wheels at a lower speed belted to dynamos at a higher speed; but in some instances, as at the lighting plant of Spokane, Wash., wheels of a higher speed are belted to dynamos of a lower speed. Another plan by which moderate dynamo speeds are obtained with water-wheels under rather high heads mounts a dynamo at each end of the shaft of a large turbine or pair of turbines. This plan is followed at the plant of the Royal Aluminum Company, Shawinigan Falls, Quebec, where there are two pairs of horizontal turbine wheels, each pair developing 3,200 horse-power under a head of 125 feet, and driving a dynamo direct-coupled on each end of its shaft. Where vertical wheels are employed it is sometimes more desirable to drive some standard type of dynamo with horizontal shaft by means of bevel gears than to design a special dynamo to mount directly on the vertical shaft. This latter plan is warranted in very large work like that at two of the Niagara Falls generating stations, where the twenty-one 3,750-kilowatt dynamos are direct-connected, each on the vertical shaft of a turbine. This type of connection is not one that will be frequently followed, but at one other point—Portland, Ore.—each dynamo is mounted directly on the shaft of its vertical turbine wheel.

Where water-wheels must operate under heads of several hundred feet, it is usually necessary to abandon pressure turbines and to adopt one of the types of impulse-wheels. In this class of wheels the peripheral speed of highest efficiency is only one-half the spouting velocity of the water under any particular head. This gives the impulse-wheels about two-thirds the peripheral speed of pressure turbines of equal diameter and consequently about two-thirds as many revolutions per minute. But as the water may be applied at one or more points on the circumference of an impulse-wheel, as desired, such wheels may have much greater diameters than pressure turbines for equal power under a given head.

FIG. 43.—Generators in Power-station at Mechanicsville on the Hudson River.

FIG. 44.—Generators at Chamblay, Quebec, Power-house.

These properties of low peripheral speed, as to head and great diameter, as to power developed, fit impulse-wheels for direct-connection to dynamos where great heads of water must be employed, and they are generally used in such cases. This is particularly true for the Pacific coast, where water-powers depend more on great heads than on large volumes. In the generating plant of the Bay Counties' Power Company, at Colgate, Cal., the dynamos are direct-connected to impulse-wheels that operate under a head of 700 feet. The three 2,250-kilowatt dynamos in this plant are each mounted on a wheel shaft operating at 285 revolutions per minute, and each of the four 1,125-kilowatt dynamos is direct-driven by an impulse-wheel at 400 revolutions per minute. At the Electra, Cal., plant of the Standard Electric Company the impulse-wheels operate at 240 revolutions per minute under a head of 1,450 feet. Each of the five pairs of these wheels drives a 2,000-kilowatt generator, direct-connected. As the head of water at these wheels is 1,450 feet, its spouting velocity is about 300 feet per second, or 18,000 feet per minute. Each wheel is eleven feet in diameter, so that a speed of 240 revolutions per minute gives the periphery a little less than 9,000 feet per minute, or about one-half of the spouting velocity of the water. These two great plants are excellent illustrations of the way in which impulse-wheels, under great heads, may be given speeds that are suitable for direct-connected dynamos.

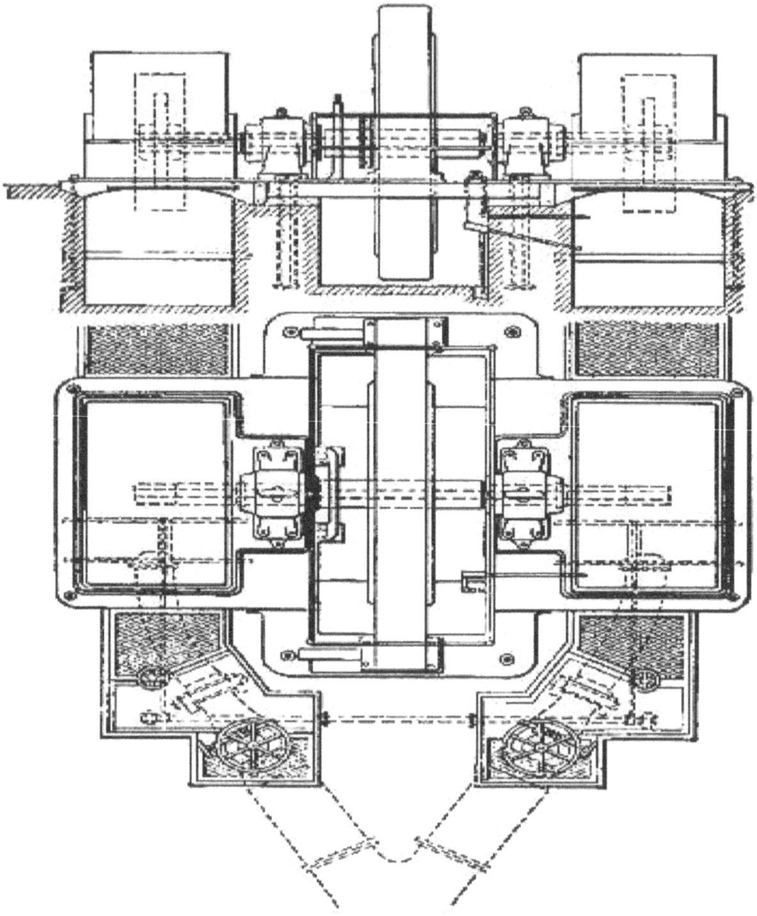

FIG. 51*a*. Plan and Elevation of Water Wheels and Generators at Power Station on Burrard Inlet, near Vancouver, B. C.

Three types of alternators, the revolving armature, the revolving magnet, and the inductor, are used in the generating plants of electric transmission systems.

Revolving armatures are used in the dynamos of comparatively few transmission systems and hardly at all in those of recent date. The prevailing type of alternator for transmission work is that with internal revolving magnets and external stationary armature. This type is employed in the great water-power plants at Cañon Ferry, Mont.; Sault Ste. Marie, Mich., and for all of the generators installed in the later Niagara Falls plants. For the sixteen earlier vertical generators at Niagara Falls the revolving magnets are external to the stationary armatures, but this construction has the disadvantage of high first cost and inaccessibility of the internal armature, and is not likely to be often adopted elsewhere.

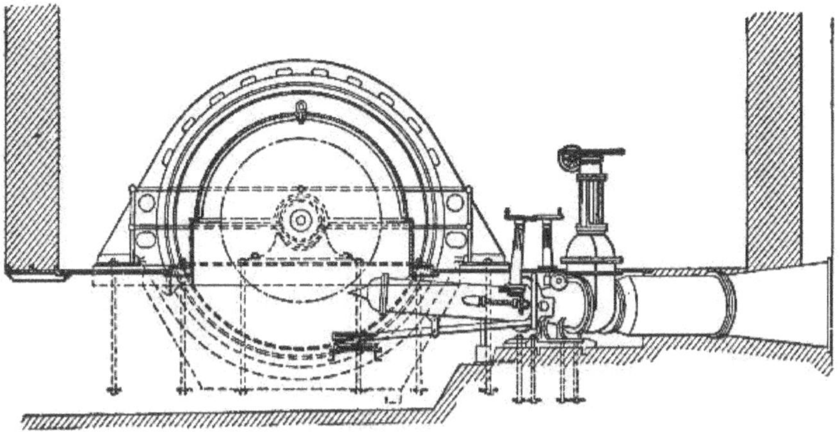

FIG. 46.—Elevations of Water-wheels and Generators at Power-station on Burrard Inlet, near Vancouver, B. C.

Inductor alternators are those in which both the armature and magnet coils are stationary and only a suitable structure of iron revolves; they are employed in a comparatively small number of transmission systems, but this number includes some of the largest plants. The seven alternators in the Colgate, Cal., plant aggregating 11,250 kilowatts capacity, and the five alternators in the plant at Electra in the same State, with a capacity of 10,000 kilowatts, are all of the inductor type. As more commonly constructed the magnet winding of the inductor alternator consists of only one or two very large coils, which are in some cases as much as ten feet in diameter. The repair of these large magnet coils seems to present a more serious problem, in case of accident, than the repair of the small coils used on interval, revolving magnets. As far as satisfactory operating qualities are concerned, inductor alternators and those with revolving magnets seem to be on an equality, but for structural reasons inductor alternators will probably be built less freely in the future than in the past.

Nearly all long transmissions are now carried out with either two- or three-phase current. The most notable two-phase installation is that at Niagara Falls, where the original ten generators, as well as the eleven dynamos later added in two of the large plants, are all of the two-phase type. At Cañon Ferry, Mont., the first four of the 750-kilowatt generators were two-phase, but the six machines of like capacity installed later are three-phase. In the latest plants of large capacity or involving very long transmissions three-phase machines have been generally employed. This is true of the Colgate and Electra plants in California, and of that at Sault Ste. Marie, Mich.

FIG. 47.—Interior of Power-house at Garvin's Falls on the Merrimac River.

FIG. 48.—500-Kilowatt Generator in Station at Great Falls on the Presumpscot River.

As to frequency, existing practice extends all the way from 133 cycles per second on the lines at Marysville, Cal., down to only 15 cycles on the transmission for the Washington & Baltimore Electric Railway.

More common practice ranges between 25 and 60 cycles. Niagara Falls saw the first great plant installed for 25 cycles, but others of that frequency are now engaged in the supply of light and power for general distribution. For transmission to electric railway lines a frequency of 25 cycles has been and is being widely used, prominent examples of which may be seen in the New Hampshire traction, the Berkshire, and the Albany & Hudson systems.

FIG. 49.—Columbus, Ga., Water-power Station.

The strong feature of a system at 25 cycles is that it is well suited to the supply of continuous currents through rotary converters with reasonable numbers of poles, armature slots, and commutator bars.

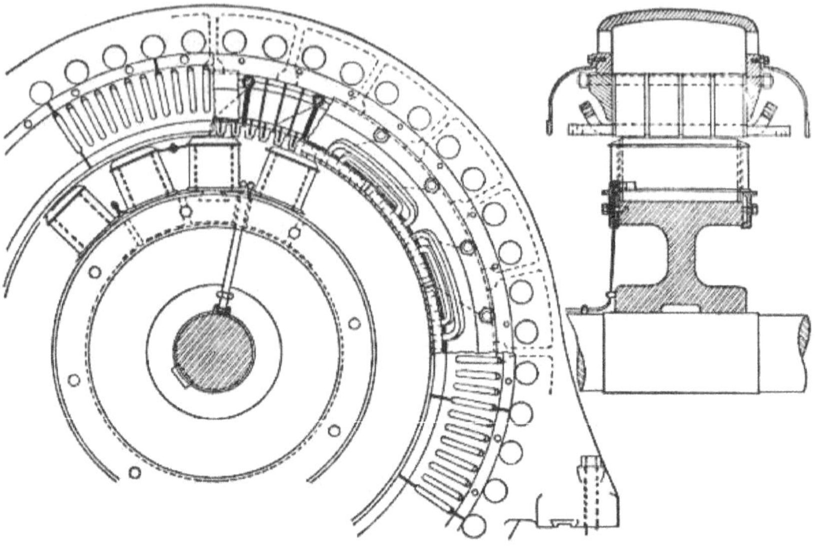

FIG. 50.—1065-Kilowatt, 2300 volt Generator Connected to Motor in Shawinigan Sub-station at Montreal.

On the other hand, the cost of transformers is greater with current at 25 cycles per second than with a higher frequency, and this current is only just bearable for incandescent lighting and quite unsuited for arc lamps, because of the fluctuating character of the light produced. At 15 cycles per second a current can be employed for incandescent lighting with satisfactory results only by means of some special devices, as lamps with very thick filaments, to avoid the flicker. Very low fluctuations cut down undesirable effects in the way of inductance and resonance, but these effects can be avoided to a large degree in other ways.

Where power is the most important element in the service of an electric water-power and transmission system there is a decided tendency to adopt a rather small number of periods for the system, even at some disadvantage as to lighting facilities. This is illustrated by the transmission from St. Anthony's Falls, Minn., at 35 cycles, from Cañon City to Cripple Creek, Col., at 30 cycles, by the Sault Ste. Marie plant of 32,000 kilowatts at 30 cycles, as well as by the two Niagara Falls plants of 78,750 kilowatts at 25 cycles.

Where the main purpose of a transmission system is the supply of light and power for general distribution, sixty periods per second are adopted as the standard in many cases. This number of periods in comparison with a smaller one tends to increase the cost of rotary converters but decreases the cost of transformers, and is suitable for both incandescent and arc lighting.

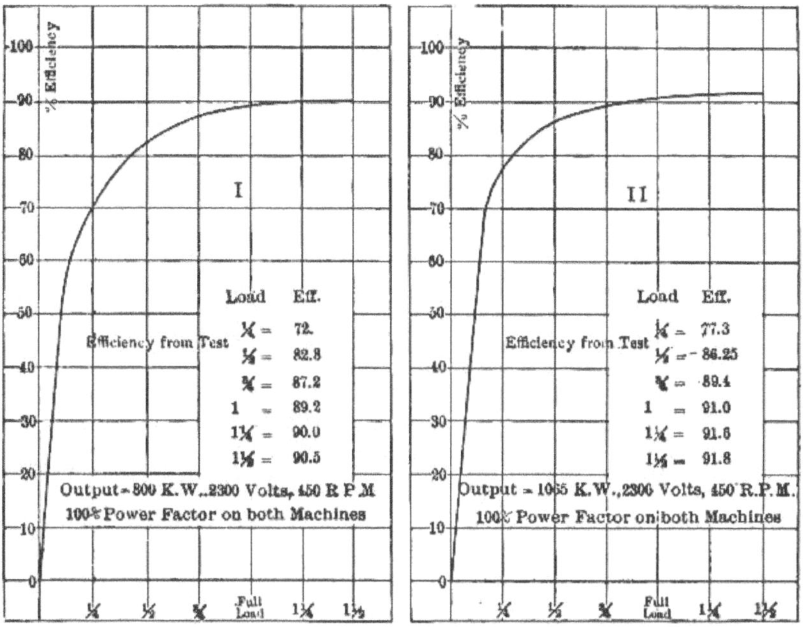

FIG. 51.—Efficiency Curves for Motor Generators at Montreal Sub-station of the Shawinigan Transmission Line.

Few, if any, transmission systems have recently been installed for frequencies above sixty cycles, and the older plants that worked at higher figures have in most cases been remodelled.

During the past decade the voltages of alternators have been greatly increased, but have not caught up with the demand for high pressures on long-transmission lines. Ten years ago when the first long transmissions were going into operation 2,000 volts was considered high for an alternator. As this voltage is too low for economy of conductors longer than three or four miles, the important early transmissions were all carried out with the aid of step-up transformers at generating stations. The practice then was, and to a large extent still is, to design the alternators for a transmission with a voltage well suited to their economical construction, and then give the step-up transformers any ratio necessary to attain the required line voltage.

Alternators in Transmission Systems.

Location of System.	Number at Plant.	Kilowatts Each.	Alternator Voltage.	Phase.	Cycles.	R. P. M.	Type of Magnet.	Method of Connections.
Niagara Falls[A]	16	3,750	2,300	2	25	250	External revolving	Direct
Niagara Falls[A]	5	3,750	2,300	2	25	250	Internal	"
Colgate to Oakland	3	2,250	2,400	3	60	285	Inductor	"
Colgate to Oakland	4	1,125	2,400	3	60	400	"	"
Electra to S.Francisco	5	2,000	....	3	60	240	"	"
Portsmouth to Pelh'm	1	2,000	13,200	3	25	83 .3	Internal	"
Portsmouth to Pelh'm	2	1,000	13,200	3	25	94	"	"
Virginia City	2	750	500	3	60	400	External	"
Ogden & Salt Lake	5	750	2,300	3	60	300	Internal	"
Chaudière Falls	2	750	10,500	3	66 .6	400	"	"
Yadkin River Falls	2	750	12,000	3	66	166	"	"
Lewiston, Me.	3	750	10,000	3	60	180	"	"
Farmington River to	2	750	500	3	60	...	"	"
Hartford, Conn.	2	600	500	2	60	...	"	"
Cañon Ferry to Butte	10	750	500	3	60	157	"	"
Apple Riv. to St. Paul	4	750	800	3	60	300	"	"
Edison Co., L. Angeles	4	700	750	3	50	...	"	"
Madrid to Bland	2	600	605	3	60	90	"	"
Cañon City to Cripple Creek	3	450	500	3	30	...	.....	"
Sault Ste. Marie	80	400	2,400	3	30	180	"	"
St. Hyacinthe, Que.	3	180	2,500	3	60	600	"	"
Great Falls to Portland, Me.	4	500	10,000	3	60	225	"	"

[A] Niagara Falls Power Company.

Thus in the two water-power plants connected with the electrical supply system of Hartford, Conn., the alternators operate at 500 volts with transformers that put the line voltage up to 10,000. In the station on Apple River that supplies the lighting system of St. Paul, Minn., the alternators operate at 800 volts, and this is raised to 25,000 volts for the line. At Cañon Ferry the alternator voltage of 500 is multiplied by 100 in the transformers giving 50,000 on the line.

FIG. 52.—Transmission Line of New Hampshire Traction Company.

Where the generating station of a transmission system is located close to a part of its load the alternators are given a voltage suitable for distribution, say about 2,400, and any desired pressure on the line is then obtained by means of step-up transformers. Two of the Niagara Falls plants are an illustration of this practice, the voltage of all the alternators there being 2,200, which is raised to 22,000 for the transmission of a part of the energy to Buffalo. A similar practice is followed in the water-power plant at Ogden, where the generators furnish current at 2,300 volts for local distribution, and transformers raise the pressure to 26,000 volts for the transmission to Salt Lake City. In the 32,000-kilowatt plant at Sault Ste. Marie, Mich., the alternators operate at 2,400 volts and a large part of their load is local, but this voltage will no doubt be raised by transformers when transmission lines are operated.

For generating stations that carry little or no local loads the cost of transformers can be saved if the generators develop the voltage required on the transmission lines. This possible saving has led to the development of alternators that generate voltages as high as 15,000 in their armature coils. Such alternators have stationary armatures in all cases and are of either the revolving magnet or inductor type.

At the present time many transmission systems in the United States operating at 10,000 or more volts develop these pressures in the armature coils of their alternators, and the number of such systems is rapidly increasing. It is now the rule rather than the exception to dispense with step-up transformers on new work where the line voltage is anything under 15,000. Perhaps the longest transmission line now in regular operation with current from the armature coils of an alternator is that at 13,200 volts between the generating station at Portsmouth and one of the sub-stations of the New Hampshire Traction system at Pelham, a distance of forty-two miles.

In at least one transmission system now under construction, that of the Washington, Baltimore & Annapolis Electric Railway, the voltage of generators to supply the line without the intervention of step-up transformers will be 15,000.

The company making these alternators is said to be ready to supply others that generate 20,000 volts in the armature coils whenever the demand for them is made. In quite a number of cases alternators of about 13,000 volts have been installed for transmissions along electric railway lines.

Systems Using High-voltage Alternators.	Alternator Voltages.
Electrical Development Co. of Ontario, Niagara Falls	12,000
Lighting and Street Railway, Manchester, N. H.	10,000
Lighting and Street Railway, Manchester, N. H.	12,500
Lighting and Power, Portland, Me.	10,000
Lighting and Power, North Gorham, Me.	10,000
Mallison Power Co., Westbrook, Me.	10,000
Lighting and Power, Lewiston, Me.	10,000
Electric Railway, Portsmouth, N. H.	13,200
Electric Railway, Pittsfield, Mass.	12,500

Ludlow Mills, Ludlow, Mass.	13,200
Electric Railway, Boston to Worcester, Mass.	13,200
Electric Railway, Albany & Hudson, N. Y.	12,000
Empire State Power Co., Amsterdam, N. Y.	12,000
Lehigh Power Co., Easton, Pa.	12,000
Hudson River Power Co., Mechanicsville, N. Y.	12,000
Light and Power, Anderson, S. C.	11,000
Fries Mfg. Co., Salem, N. C.	12,000
Light and Power, Ouray, Col.	12,000
Washington & Baltimore Electric Railway	15,000
Canadian Niagara Power Co., Niagara Falls	12,000
Ontario Power Co., Niagara Falls	12,000

This list of high-voltage alternators is not intended to be exhaustive, but serves to indicate their wide application. If such alternators can be purchased at a lower price per unit of capacity than alternators of low voltage plus step-up transformers, there is an apparent advantage for transmission systems in the high-voltage machines. This advantage may rest in part on a higher efficiency in the alternators that yield the line voltage than in the combination of low-voltage alternators plus step-up transformers. It is not certain, however, that depreciation and repairs on the generators of high voltage will not be materially greater than the like charges on generators of low voltage, and some advantage in price should be required to cover this contingency.

Just how far up the voltage of alternators can be pushed for practical purposes is uncertain, but it seems that the limit must be much below that for transformers where there is ample room for solid insulation and the coils can be immersed in oil. The use of generators at 10,000 volts and above tends to lower the volts per mile on transmission lines, because it seems better in some cases to increase the weight of line conductors rather than to add step-up transformers, as in the 42-mile transmission from Portsmouth to Pelham.

CHAPTER X.
TRANSFORMERS IN TRANSMISSION SYSTEMS.

Transformers are almost always necessary in long electric systems of transmission, because the line voltage is greater than that of generators, or at least that of distribution. As transformers at either generating or receiving stations represent an increase of investment without corresponding increase of working capacity, and also an additional loss in operation, it is desirable to avoid their use as far as is practicable. In short transmissions over distances of less than fifteen miles it is generally better to avoid the use of transformers at generating stations, and in some of these cases, where the transmission is only two or three miles, it is even more economical to omit transformers at the sub-stations.

Thus, where energy is to be transmitted two miles and then applied to large motors in a factory, or distributed at 2,500 volts, the cost of bare copper conductors for the three-phase transmission line will be only about $6 per kilowatt of line capacity at 2,500 volts, with copper at 15 cents per pound, and a loss of 5 per cent at full load. The average loss in such a line will probably be as small as that in one set of transformers and a line of higher voltage. Furthermore, the first cost of the 2,500-volt generators and line without transformers will be less than that of generators and line of higher voltage with step-down transformers at the sub-station.

As generators up to 13,500 volts are now regularly manufactured, it is quite common to omit step-up transformers at the main stations of rather short transmission systems. This practice was followed in the 13,500-volt transmission to Manchester, N. H., the 10,000-volt transmission to Lewiston, Me., and the 12,000-volt transmission to Salem, N. C.

In most transmission over distances of twenty-five miles or more, step-up transformers at generating stations as well as step-down transformers at sub-stations are employed. As yet the highest voltages that have been put into practical use on transmission lines (that is, 50,000 to 60,000) are much below the pressures that have been yielded by transformers in experimental work. These latter voltages have in a number of instances gone above 100,000. The numbers and capacities of transformers used at main stations vary much in their relation to the numbers and individual capacities of generators there. In some cases there are three times as many transformers as three-phase generators, and the capacity of each transformer is either equal to or somewhat greater than one-third of the capacity of each generator.

FIG. 53.—Transformers at Central Sub-station, Montreal.

Thus in the station at Spier Falls on the Hudson, whence power is transmitted to Albany and other cities, the number of step-up transformers will be thirty and their aggregate capacity will be 24,014 kilowatts, while the total number of three-phase generators will be ten, with a combined capacity of 24,000 kilowatts. Another practice is to give each transformer a capacity greater than one-third of that of the three-phase generator with which it is to be connected, and make the total number of transformers less than three times as great as the number of generators. An example of this sort exists in the station on Apple River, whence power is transmitted to St. Paul. This station contains four three-phase generators of 750 kilowatts each, and six transformers of 500 kilowatts each, these latter being connected in two sets of three each. The use of three transformers for each three-phase generator instead of three transformers for each two or three generators, tends to keep transformers fully loaded when in use, and therefore to increase their efficiency. On the other hand, efficiency increases a little with the size of transformers, and the first cost per unit capacity is apt to be less the greater the size of each.

Another solution of the problem is to provide one transformer for each three-phase generator, each transformer being wound with three sets of coils, so that the entire output of a generator can be sent into it. This practice is followed at the Hochfelden water-power station, whence power is transmitted to Oerlikon, Switzerland, also in the water-power station at Grenoble, France, whence energy at 26,000 volts is transmitted to a number of factories. With three-phase transformers each generator and its transformer may form an independent unit that can be connected with the line at pleasure, thus tending to keep transformers at full load.

Though three-phase transformers are much used in Europe, they have thus far had little application in the United States. Single-phase transformers may, of course, be limited in number to that of the three-phase generators with which they are used, but such transformers must regularly be connected to the generators and line in groups of two or three. Such an equipment was provided in part at the 7,500-kilowatt station on the Missouri River at Cañon Ferry, which contains ten three-phase generators of 750 kilowatts each. The transformers at this station include twelve of 325 kilowatts each, connected in four groups of three each, also six transformers of 950 kilowatts each which are also connected in groups of three. Three of these larger transformers have a capacity of 2,850 kilowatts, or nearly equal to that of four generators.

With two-phase generators single-phase transformers must be connected in pairs, and it is common to provide two transformers for each generator. Thus, in the Rainbow station on the Farmington River, whence energy is transmitted to Hartford, there are two generators of the two-phase type and rated at 600 kilowatts each, also four transformers rated at 300 kilowatts each.

As the regulation of transformers on overloads is not as good as that of generators, it seems good practice to give each group of transformers a somewhat greater capacity than that of the generator or generators whose energy is to pass through it. This plan was apparently followed at the Cañon Ferry station, where the total generator capacity is 7,500 kilowatts and the total capacity of step-up transformers is 9,600 kilowatts. Each group of the 325-kilowatt transformers there has a capacity of 975 kilowatts, while each generator is only of 750 kilowatts. Usually the number of groups of transformers at a two-phase or three-phase generating station is made greater than the number of transmission circuits supplied by the station, for some of the reasons just considered. When this is not the case it is commonly desirable in any event to have as many groups of step-up transformers as there are transmission circuits, so that each circuit may be operated with transformers that are independent of the other circuits.

At sub-stations it is desirable to have a group of transformers for each transmission circuit, and it may be necessary to subdivide the transformer capacity still further in order to keep transformers in operation at nearly full load, or to provide a group of transformers for each sort of service or for each distribution circuit. All of the transformers at a sub-station should have a total capacity at least equal to that of the generators whose energy they are to receive, minus the losses in step-up transformers and the line. Transformers at sub-stations do not necessarily correspond in number or individual capacity with those at generating stations, and the number of sub-station transformers bears no necessary relation to the number of generators by which they are fed.

Two transmission circuits extend from Cañon Ferry to a sub-station at Butte, and in that sub-station there are six transformers divided into two groups for three-phase operation, each transformer being rated at 950 kilowatts. This sub-station equipment thus corresponds to only the six 950-kilowatt transformers in the generating station, because the four groups of smaller transformers there are used to supply the transmission line to Helena.

In the sub-station at St. Paul that receives the entire output of the plant on Apple River, where the six transformers of 500 kilowatts each are located, ten transformers receive energy from two three-phase transmission circuits. Six of these transformers are rated at 300 kilowatts each. The 300-kilowatt transformers are connected in two groups of three each, and the 200-kilowatt in two groups of two each, transforming current from three-phase to two-phase. The aggregate capacity of the sub-station transformers is thus 2,600 kilowatts, while that of transformers at the generating station is 3,000 kilowatts. With four generators at the water-power plant there are ten transformers at the sub-station, where all the energy, minus losses, is delivered.

At Watervliet, where one of the several sub-stations of the system with its larger generating plant at Spier Falls is located, the capacity of each transformer is 1,000 kilowatts, though each transformer at Spier Falls has a rating below this figure.

In the sub-station at Manchester, N. H., that receives nearly all of the energy from four water-power plants, containing eight generators with an aggregate capacity of 4,030 kilowatts, there are located twenty-one step-down transformers that have a total rating of 4,200 kilowatts. These twenty-one transformers are fed by six circuits, of which five are three-phase and one is two-phase. A part of the transformers supply current to motor-generators, developing 500-volt current for a street railway, and the remaining transformers feed circuits that distribute alternating current.

From these examples it may be seen that in practice either one or more groups of transformers are employed in sub-stations for each transmission circuit, that the total number of these transformers may be just equal to or several times that of the generators from which they receive energy, and that the individual capacities of the transformers range from less than one-third to more than that of a single generator. Groups of transformers at a main station must correspond in voltage with that of the generators in the primary and that of the transmission line in the secondary windings. Sub-station transformers receive current at the line voltage and deliver it at any of the pressures desired for local distribution. Where step-up transformers are employed the generator pressure in nearly all cases is at some point between 500 and 2,500 volts.

At the Cañon Ferry station the voltage of transformers is 550 in in the primary and 50,000 in the secondary windings. In the Colgate power-house, whence energy is transmitted to Oakland, the generator pressure of 2,400 volts is raised to 40,000 volts by transformers. Generator voltage in the power-house on Apple River is 800 and transformers put the pressure up to 25,000 for the line to St. Paul. Transformers at the Niagara Falls station raise the voltage from 2,200 to 22,000 for the transmission to Buffalo.

As transformers can be wound for any desired ratio of voltages in their primary and secondary coils, a generator pressure that will allow the most economical construction can be selected where step-up transformers are employed. In general it may be said that the greater the capacity of each generator, the higher should be its voltage and that of the primary coils of step-up transformers, for economical construction. At sub-stations the requirements of distribution must obviously fix the secondary voltages of transformers.

Weight and cost of transformers depend in part on the frequency of the alternating current employed, transformers being lighter and cheaper the higher the number of cycles completed per second by their current, other factors remaining constant. In spite of this fact the tendency during some years has been toward lower frequencies, because the lower frequencies present marked advantages as to inductive effects in transmission systems, the distribution of power through induction motors, the construction and operation of rotary converters, and the construction of generators. Instead of the 133 cycles per second that were common in alternating systems when long transmissions first became important, sixty cycles per second is now the most general rate of current changes in such transmission systems. But practice is constantly extending to still lower frequencies. The first Niagara Falls plant with its twenty-five cycles per second reached the lower limit for general distribution, because incandescent lighting is barely satisfactory and arc lighting decidedly undesirable at this figure.

In contrast with the great transmissions from Cañon Ferry to Butte, Colgate to Oakland, and Electra to San Francisco, which operate at sixty cycles, the system between Cañon City and Cripple Creek, in Colorado, as well as the great plant at Sault Ste. Marie, employs thirty-cycle current, and the lines from Spier Falls to Schenectady, Albany, and Troy are intended for current at forty cycles per second. From these examples it may be seen that the bulk and cost of transformers is not the controlling factor in the selection of current frequency in a transmission system.

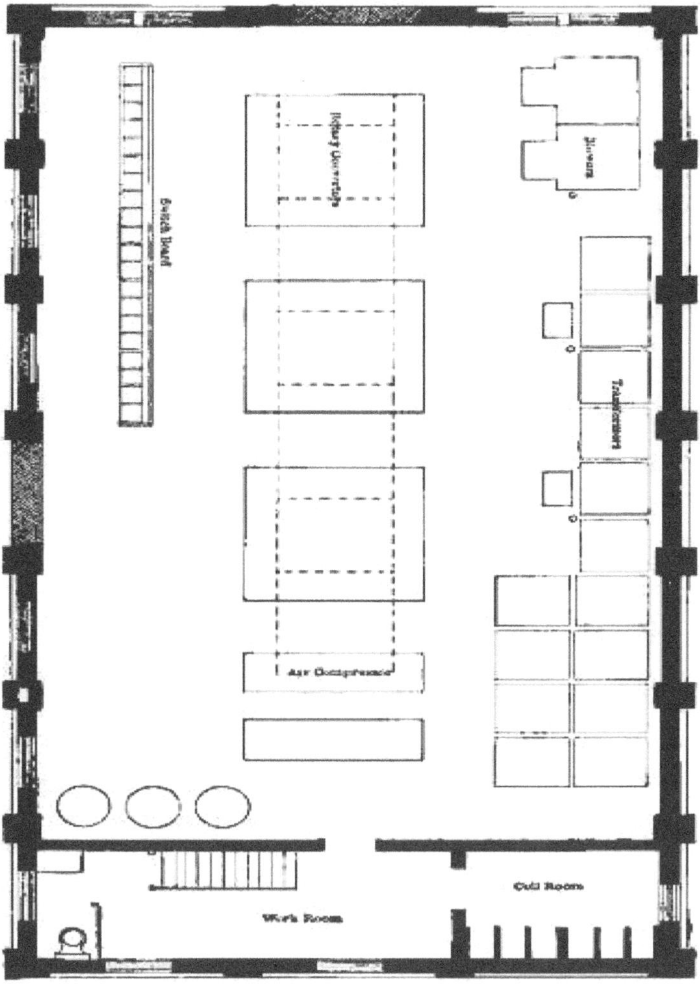

FIG. 54.—First Floor of Saratoga Sub-station.

Transformers used at either generating or sub-stations are cooled by special means in many cases.

The advantages of so-called artificial cooling are smaller weight and first cost in transformers, and perhaps longer life for the insulation of windings. For these advantages a small increase in the cost of operation must be paid. Station transformers are usually cooled either by forcing air through their cases under pressure, or else by passing water through pipes in the oil with which the transformer cases are filled. If cooling with air-blast is adopted, a blower, with electric motor or some other source of power to operate it, must be provided. Where transformers are oil-insulated and cooled with water there must be some pressure to maintain the circulation. If free water under a suitable head can be had for the cooling of transformers, as in most water-power plants, the cost is very slight. Where water must be purchased and pumped through the transformers its cost will usually be greater than that of cooling with air-blast. One manufacturer gives the following as approximate figures for the rate at which water at the temperature of 15° centigrade must be forced through his transformers to prevent a rise of more than 35° centigrade in their temperature, probably when operating under full loads.

Trans-formers—Kilowatts.	Gallons per minute.
150	0.5
400	.75
400	1.00
1,000	1.5
75	.37

An air-blast to cool transformers at main or sub-stations may be provided in either of two ways. One plan is to construct an air-tight compartment, locate the transformers over openings in its top, and maintain a pressure in the compartment by means of blower-fans that draw cool air from outside. Such an arrangement has been carried out at the sub-station in Manchester, N. H. The basement underneath this sub-station is air-tight, and in the concrete floor over it there are twenty-seven rectangular openings, each twenty-five by thirty inches, and intended for the location of a 200-kilowatt transformer. Aggregate transformer capacity over these openings will thus be 5,400 kilowatts. Pressure in this basement is maintained by drawing outside air through a metal duct that terminates in a hood on the outside of the sub-station about nine feet above the ground. In the roof of this sub-station there are ample skylight openings to permit the exit of hot air that has been forced through the transformers. In the air-tight basement are two electric motors of ten horse-power each, connected to the blower that maintains the pressure. It may be noted that in this case there is less than one-horse power of motor capacity for each 200 kilowatts capacity in transformers.

Where there are not more than six or nine transformers to be cooled, it is common practice to provide a separate motor and blower for each group of three transformers, and lead the air directly from each blower to its group of transformers by a metal duct, thus avoiding the necessity for an air-chamber. In such cases a blower giving a three-eighth-ounce air pressure per square inch and a motor of one horse-power capacity are generally provided for each group of three transformers rated at 100 to 150 kilowatts each. Where cooling with air-blast is adopted, oil-insulation cannot be carried out because the air must come into intimate contact with the transformer coils and core. Both oil-insulation with water cooling and dry insulation with cooling by air-blast have been widely used in transmission systems of large capacity and high voltage.

In the Colgate plant, where the line pressure is 40,000 volts, the 700-kilowatt transformers are oil-insulated and water-cooled, and this is also true of the 950-kilowatt transformers in the 50,000-volt transmission between Cañon Ferry and Butte. On the other hand, the transmission system between Spier Falls, Schenectady, and Albany, carried out at 26,500 volts, includes transformers that range from several hundred to 1,000 kilowatts each in capacity and are all air-cooled. Either a water-cooled transformer or one cooled by air-blast may be safely overloaded to some extent, if the circulation of air or water is so increased that the overload does not cause heating beyond the allowable temperature.

The circulation of air or water through a transformer should never be forced to an extent that cools the transformer below the temperature of the air in the room where it is located, as this will cause the condensation of water on its parts.

In some cases it is desirable that means for the regulation of transformer voltages through a range of ten per cent or more each way from the normal be provided. This result is reached by the connection of a number of sections at one end of the transformer winding to a terminal board, where they may be cut in or out of action at will. Regulation is usually desired, if at all, in a secondary winding of comparatively low voltage, and the regulating sections generally form a part of such winding, but these sections may be located in the primary winding.

In order to keep the number of transformers smaller and the capacity of each larger than it would otherwise be, it is practicable to divide the low-voltage secondary winding of each transformer into two or more parts that have no electrical connection with each other. These different parts of the winding may then be connected to distinct distribution lines or other services. An example of this sort exists in the Hooksett sub-station of the Manchester, N. H., transmission system. Three-phase current at about 11,000 volts enters the primary windings of three transformers at this sub-station. Each of these transformers has a single primary, but two distinct secondary windings. Three of these secondaries, one on each transformer, are connected together and feed a rotary converter at about 380 volts, three-phase. The other three secondary windings are connected in like manner to a second rotary converter. Each of these transformers is rated at 250 kilowatts, and each rotary is rated at 300 kilowatts, so that the transformer capacity amounts to 750 kilowatts and that of the converters to 600 kilowatts, giving a desirable margin of transformer capacity for railway service. With the ordinary method of connection and windings, six transformers of 125 kilowatts each would have been required in this sub-station.

High voltage for transmission lines may be obtained by the combination of two or more transformers with their secondary coils in series. This method was followed in some of the early transmissions, as in that at 10,000 volts to San Bernardino and Pomona, begun in 1891, where twenty transformers, giving 500 volts each, were used with their high-voltage coils in series. Some disadvantages of such an arrangement are its high cost per unit of transformer capacity and its low efficiency.

In a single-phase system the maximum line pressure must be developed or received in the coils of each transformer, unless two or more are connected in series. This is also true as to either phase of a two-phase system with independent circuits. In three-phase circuits the coils of a transformer connected between either two wires obviously operate at the full line pressure. The same result is reached when the three transformers of a group are joined to the line in mesh or Δ-fashion. If the three transformers of a group are joined in star or Y-fashion, the coils of each transformer are subject to fifty-eight per cent of the voltage between any two wires of the three-phase line on which the group is connected. It is no longer the practice to connect two or more transformers in series either between two wires of a two-phase or between two wires of a three-phase circuit, because it is cheaper and more efficient to use a single transformer in each of these positions. Where very high voltage must be developed or received with a three-phase system, the star or Y-connection of each group of three transformers has the advantage of a lower strain on the insulation of each transformer than that with the mesh or Δ-grouping. Thus if the Δ-grouping is used, the line pressure equals that of each transformer coil, but if the Y-grouping is used the line voltage is 1.73 times that of each transformer coil.

At the Colgate power-house, the 700-kilowatt transformers are designed for a maximum pressure of 60,000 volts on the three-phase line when Y-connected, so that the corresponding voltage is 34,675 in their secondary coils. The primary coils of these same transformers are connected in mesh or Δ-form and each coil operates at 2,300 volts, the generator pressure.

Transformers are in some cases provided with several sets of connections to their coils so that they may be operated at widely different pressures. Thus, in the Colgate plant, each transformer has taps brought out from its secondary coils so that it can be operated at either 23,175, 28,925, or 34,675, with 2,300 volts at its primary coil. Corresponding to the three voltages named in each secondary coil are voltages of 40,000, 50,000, and 60,000 on a three-phase line connected with three of these transformers in Y-fashion.

The mesh or Δ-connection is used between the coils of transformers on some transmission lines of very high voltage. The 950 kilowatt transformers in the system between Cañon Ferry and Butte illustrate this practice, being connected Δ-fashion to the 50,000-volt line.

When transformers that will operate at the desired line voltage on Δ-connection can be obtained at slight advance over the cost of transformers requiring Y-connections, it is often better practice to select the former, because this will enable an increase of seventy-three per cent in the voltage of transmission to be made at any future time by simply changing to Y-connections. Such an increase of voltage may become desirable because of growing loads or extension of transmission lines.

An example of this sort came up some time ago in connection with the transmission between Ogden and Salt Lake City, which was operating at 16,000 volts, three-phase, with the high-pressure coils of transformers connected in Δ-form. By changing to Y-connections the line voltage was raised seventy-three per cent without increasing the strain on transformer insulation.

In some cases it is desirable to change alternating current from two-phase to three-phase, or *vice versa*, for purposes of transmission or distribution, and this can readily be done by means of static transformers. One method often employed to effect this result includes the use of two transformers connected to opposite phases of the two-phase circuit. The three-phase coil of one of these transformers should be designed for the desired three-phase voltage, and should have a tap brought out from its central point. The three-phase coil of the other transformer should be designed for 87 per cent of the desired three-phase voltage. One end of the coil designed for 87 per cent of the three-phase voltage should be connected to the centre tap of the three-phase coil in the other transformer. The other end of the 87 per cent coil goes to one wire of the three-phase circuit. The other two wires of this circuit should be connected, respectively, to the outside end of the coil that has the central tap. As a matter of illustration it may be required to transform 500-volt, two-phase current from generators, to 20,000-volt, three-phase current for transmission. Two transformers designed for 500 volts in their primary coils are necessary for this work. One of these transformers should have a secondary coil designed for 20,000 volts, so that the ratio of transformation is 20,000 ÷ 500 or 40 to 1, and a tap should be brought out from the centre of this coil. The other transformer should have a secondary voltage of 0.87 × 20,000 = 17,400, so that its ratio of transformation is 34.8 to 1.

These two transformers, with the connections above indicated, will change the 500-volt, two-phase current to 20,000 volts, three-phase.

At one of the water-power stations supplying energy for use in Hartford, four transformers of 300 kilowatts each change 500-volt, two-phase current from the generators to 10,000-volt, three-phase, for the transmission line.

In the Niagara water-power station the generators deliver two-phase current at 2,200 volts, and 975-kilowatt transformers are connected in pairs to change the pressure to 22,000 volts, three-phase, for transmission to Buffalo.

A transformer is used in some cases to raise the voltage and compensate for the loss in a transmission line. For this purpose the secondary of a transformer giving the number of volts by which the line pressure is to be increased is connected in series with the line. The primary winding of this transformer may be supplied from the line boosted or from another source.

Transformers ranging in capacity from 100 to 1,000 kilowatts each, such as are commonly used for transmission work, have efficiencies of 96 to 98 per cent at full loads, when of first-class construction. Efficiency increases slowly with transformer capacity within the limits named, and 98 per cent can be fairly expected in only the larger sizes. In any given transformer the efficiency may be expected to fall a little, say one or two per cent, between full load and half load, and another one per cent between half load and quarter load. These figures for efficiencies at partial loads vary somewhat with the design and make of transformers. In general, it may be said that step-up or step-down transformers will cost approximately $7.50 per kilowatt capacity, or about one-half of the like cost of low-voltage dynamos. If dynamos of voltage sufficiently high for the transmission line can be had at a figure below the combined cost of low-volt dynamos and raising transformers, it will usually pay to avoid the latter and develop the line voltage in the armature coils. This plan avoids the loss in one set of transformers.

TRANSFORMERS IN TRANSMISSION SYSTEMS.

Transmission System.	Transformers at Power-stations.		Transformers at Sub-stations.		Generators at Power-stations.	
	No.	Kw. Each.	No.	Kw. Each.	No.	Kw. Each.
Cañon Ferry to Butte	12	325	[A]	[A]	...	...
	6	950	6	950	10	750
Apple River to St. Paul	...	...	6	300	...	...
	6	500	4	200	4	750
White River to Dales	3	400	3	375	2	500
Farmington River to Hartford	4	300	...	...	2	600
Ogden to Salt Lake	[B]9	250	...	...	5	750
Colgate to Oakland	...	700	...	...	3	1125
					4	2250
Presumpscot River to Portland	...	...	6	200	...	...
			3	150	4	500
Four water-powers to Manchester	...	...	21	200	1	180
					3	300
					1	450
					4	650
					1	1200

[A] Other transformers at Helena sub-stations.
[B] Part of energy distributed directly from generators.

CHAPTER XI.
SWITCHES, FUSES, AND CIRCUIT-BREAKERS.

Electrical transmission has worked a revolution in the art of switching. As long as the distances to be covered by distribution lines required pressures of only a few hundred volts, the switch contacts for generators and feeders could well be exposed in a row on the surface of vertical marble slabs and separated from each other by distances of only a few inches. These switches were capable of manual operation even at times of heavy overload without danger of personal injury to the operator or of destructive arcing between the parts of a single switch or from one switch to another near-by. On the back of these marble slabs one or more sets of bare bus-bars could be located without much probability that an accidental contact between them would start an arc capable of destroying the entire switchboard structure and shutting down the station.

The rise of electric pressures to thousands and tens of thousands of volts in distribution and transmission systems has vastly increased the difficulty of safe and effective control with open-air switches. The higher the voltage of the circuit to be operated under load the greater must be the distance between the contact parts of each switch and also between adjacent switches. Such switches must also be farther removed from the operators as the voltages of their circuits go up, as a person cannot safely stand very close to an electric arc of several feet or even yards in length. In the West, where long transmissions are most common, long break-stick switches have been much used with high voltages. These switches depend on the length of the break to open the circuit and on the length of the stick that moves the switch-jaw or plug to insure the safety of the operator. Where switches of this sort are used it is highly important to have ample distances between the contact points of each switch and also between the several switches. On circuits of not more than 10,000 volts an arc as much as a yard long will in some cases follow the opening switch blade and hold on for several seconds. On the 33,000-volt transmission line at Los Angeles a peculiar form of switch is used which makes a break between a pair of curved wire horns that are ten inches apart at their nearest points. When the contact between these horns is broken the arc travels up between portions of the horns that curve apart and is thus finally ruptured. Besides the very large space required for open switches on circuits of 5,000 to 10,000 volts or more, there is a further objection that the arcs developed by opening such switches under heavy loads rapidly destroy the contact parts and produce large quantities of metallic vapor that is objectionable in a central station. In some experiments performed at Kalamazoo (A. I. E. E., vol. xviii., p. 407) with open-air switches the voltages ranged from 25,000 to 40,000. The loads on circuits broken by the switches were highly inductive and mounted from 1,200 to 1,300 kilovolt-amperes. At 25,000 volts the arc produced by the open-air switch held on for several seconds. At 40,000 volts the arc following the opening of this switch was over thirty feet long, and being out of doors near the pole line the arc struck the line wires and short-circuited the system. It has been shown that the oscillations of voltage occurring when a circuit under heavy load is opened by an open-air switch may be very dangerous to insulation (A. I. E. E., vol. xviii., p. 383). In the Kalamazoo test the oscillations of this sort were reported to have reached two or three times the normal voltage of the system when the open-air switch was used.

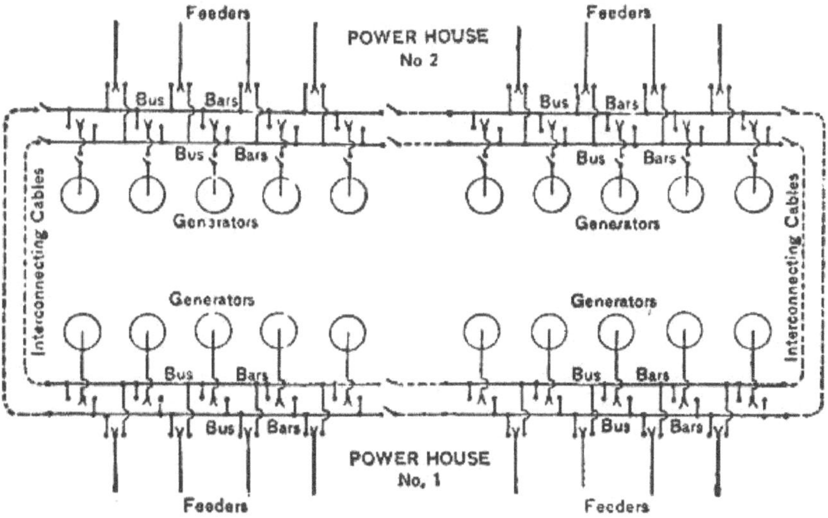

FIG. 55.—Connections between Power-houses 1 and 2 at Niagara Falls.

Facts of the nature just outlined have led to the development of oil switches. The general characteristic of oil switches is that the contact parts are immersed in, and the break between these contacts takes place under, oil. Two types of the oil switch are made, one having all of its contact parts in the same bath of oil and the other having a separate oil-bath for each contact. Compared with those of the open-air type, oil switches effect a great saving of space, develop no exposed arcs or metallic vapors, cause little if any oscillation or rise of voltage in an alternating circuit, and can be depended on to open circuits of any voltage and capacity now in use. In the tests above mentioned at Kalamazoo, a three-phase oil switch making two breaks in each phase and with all the six contacts in a single oil-bath was used to open circuits of 25,000 volts and 1,200 to 1,300 kilovolt-arcs with satisfactory results. At 40,000 volts, however, this type of switch spat fire and emitted smoke, indicating that it was working near its ultimate capacity. A three-phase switch with each of its six contacts in a separate cylindrical oil-chamber was used to open the 40,000-volt 1,300 kilovolt-arc circuit at Kalamazoo with perfect success even under conditions of short-circuit and without the appearance of fire or smoke at the switch. The three-phase switch used in the tests at Kalamazoo and having each of its contacts in a separate oil-chamber was similar in construction to the switches used in the Metropolitan and Manhattan railway stations in New York City. In each of these switches the two leads of each phase terminate in two upright brass cylinders. These cylinders have fibre linings to prevent side-jumping of the arcs when the switch is opened, and each cylinder is filled with oil. Into the two brass cylinders of each phase dips a ∩-shaped contact piece through insulating bushings, and the ends of this contact piece fit into terminals at the bottom of the oil pots. A wooden rod joins the centre or upper part of the ∩-contact piece, and the three rods of a three-phase switch pass up through the switch compartment to the operating mechanism outside. The six brass cylinders and their three ∩-contact pieces are usually mounted on a switch cell built entirely of brickwork and stone slabs. For a three-phase switch the brick and stone cell has three entirely separate compartments, and each compartment contains the two brass cylinders that form the terminals of a single phase. On top of and outside the cell the mechanism for moving the wooden switch rods is mounted. In the Metropolitan station, where the voltage is 6,000, the vertical movement of the ∩-shaped contact piece with its rod is twelve inches. At the Manhattan station, where the operating voltage is 12,000, the vertical movement of the ∩-contacts in opening a switch is seventeen inches. The total break in each phase in a switch at the Metropolitan station is thus twenty-four inches, or four inches per 1,000 volts, and the total break per phase in switches at the Manhattan station is thirty-four inches, or 2.66 inches per 1,000 volts total pressure.

Oil switches are now very generally employed on alternating circuits that operate at 2,000 volts or more for purposes of general distribution. On circuits of moderate voltage like that just named, and even higher, it is common practice to use oil switches that have only a single reservoir of oil each, the entire six contacts in the case of a three-phase switch being immersed in this single reservoir. Such switches are usually operated directly by hand and are located on the backs of or close to the slate or marble boards on which the handles that actuate the switch mechanism are located. A good example of this sort of work may be seen at the sub-station in Manchester, N. H., where energy from four water-power stations is delivered over seven transmission lines and then distributed by an even larger number of local circuits at 2,000 volts three-phase. At the Garvin's Falls station, one of the water-power plants that delivers energy to the sub-station in Manchester, the generators operate at 12,000 volts three-phase, and these generators connect directly with the bus-bars through hand-operated oil switches on the back of the marble switchboard. These last-named switches, like those at the Manchester sub-station, have all the contacts of each in a single reservoir of oil.

With very high voltages, where only a few hundred kilowatts are concerned, and also with powers running into thousands of kilowatts at as low a pressure as 2,000 volts, it is very desirable to remove even oil switches from the switchboard and the vicinity of the bus-bars. Great powers as well as very high voltages not only increase the element of personal danger to an attendant who must stand close to a switch while operating it, but also render the damage to other apparatus that may result from any failure of or short-circuit in a switch much more serious.

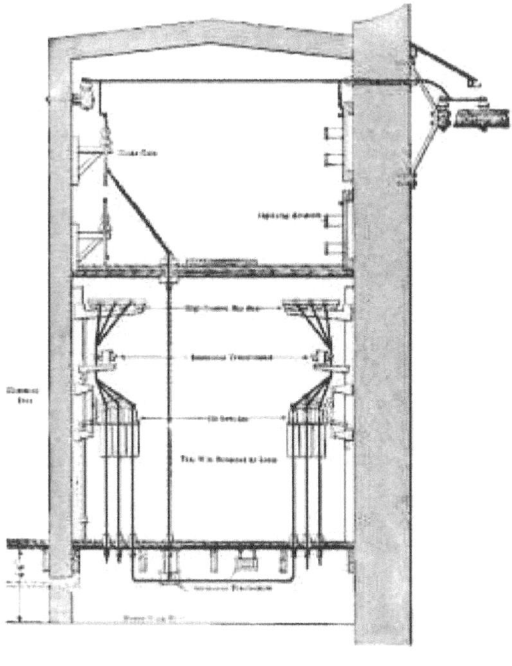

FIG. 56.—Wire-room Back of Switchboard in Power-station on French Broad River, North Carolina.

As soon as the switches are removed to a distance from the operating board the necessity for some method of power control becomes evident, since the operator at the switchboard should be able to make or break connections of any part of the apparatus quickly. The necessity for the removal of switches for very large powers to a distance from the operating boards and for the application of mechanical power to make and break connections was met before the development of oil switches. Thus at the first Niagara (A. I. E. E., vol. xviii., p. 489) power-house, in 1893, the switches for the 3,750-kilowatt, 2,200-volt generators, though of the open-air type, were located in a special switch compartment erected in the generator room and over a cable subway at some distance from the operating board. These switches were actuated through compressed-air cylinders into which air was admitted by the movement of levers near the switchboard. Evidently a switch of this capacity—1,000 amperes per pole and 2,200 volts, two-phase—could not well be operated by hand-power wherever located, because of the large effort required. In the second generating station at Niagara Falls oil switches similar to those used at the Manhattan Elevated Railway plant in New York, but two-phase, were employed. Each of these oil switches at Niagara Falls has a capacity of 5,000 horse-power, like the previous open-air switches, and is electrically actuated.

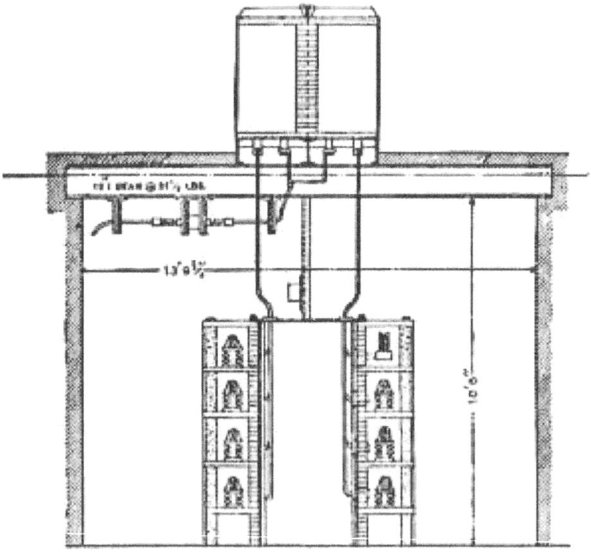

FIG. 57.—Section through Cable Subway under Oil Switches in Niagara Power-house No. 2.

In these electrically operated oil switches a small motor is located on top of the brick cell that contains the contact parts, and this motor releases and compresses springs that open and close the switch. While it is not desirable to employ open-air switches to open circuits of several thousand or even hundreds of kilowatts at voltages of 2,000 or more, it is nevertheless possible to do so. This is shown by the experience of the first Niagara Falls station, where the 2,200-volt two-phase switches are reported to have opened repeatedly currents of more than 600 amperes per phase without injurious sparking. The great rise of voltage that was shown by the experiments at Kalamazoo to follow the opening of a simple open-air switch was avoided at the first Niagara switches by a simple expedient. In these 5,000 horse-power open-air switches a shunt of high resistance was so connected between each pair of contacts that the blades and jaws that carried the main body of the current never completely opened the circuit. When the main jaws of one of these switches were opened the shunt resistance continued in circuit until subsequently broken at auxiliary terminals. That no excessive rise of voltage took place when one of these switches was open was shown by connecting two sharp terminals in parallel with the switch and by adjusting these terminals to a certain distance apart. Had the voltage risen on opening the switch above the predetermined amount there would have been an arc formed by a spark jumping the distance between the pointed terminals.

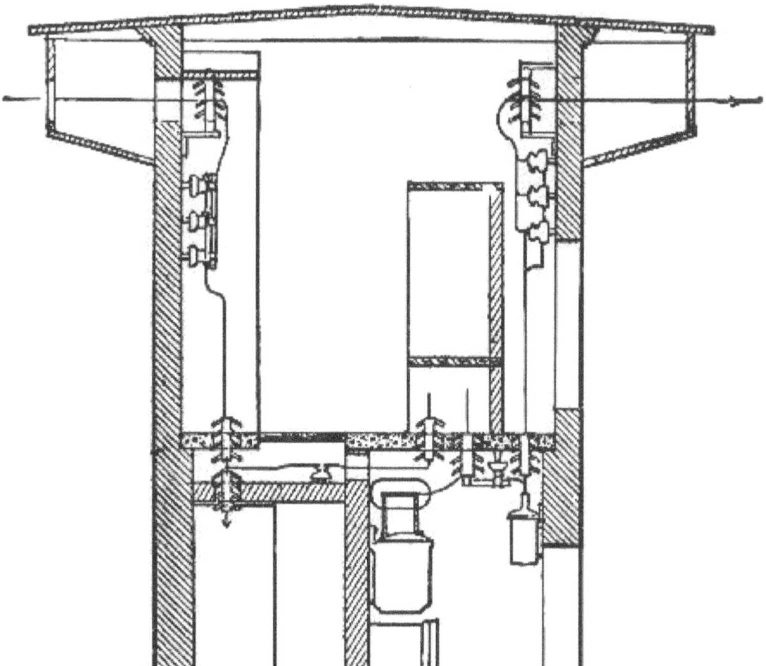

FIG. 58.—Schenectady Switch-house on Spier Falls Line.

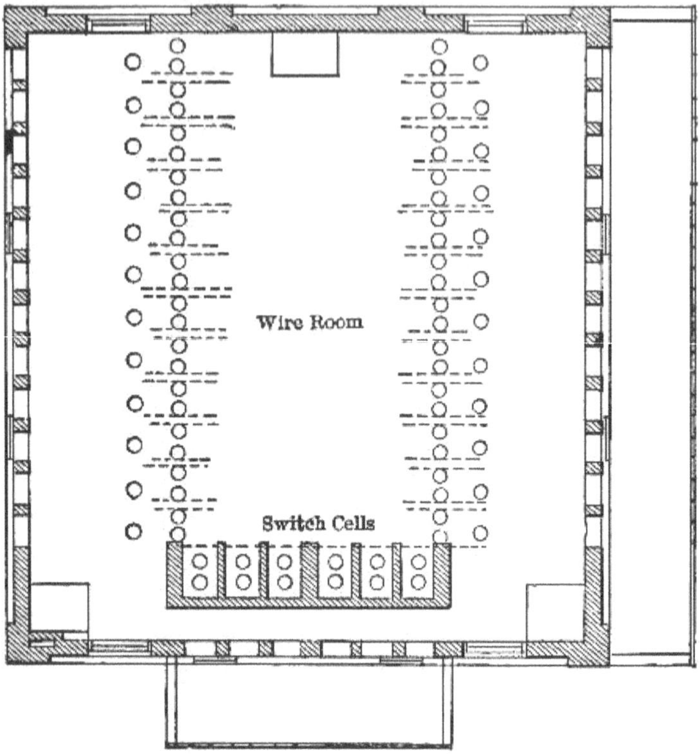

FIG. 59.—Second-floor Plan of Saratoga Switch-house on Spier Falls Line.

Safety and reliability of operation at high voltages, say of 5,000 or more, require that each element of the equipment be so isolated as well as insulated from every other element that the failure or even destruction of one element will not seriously endanger the others. With this end in view the cables from each generator to its switch should be laid in a conduit of brick or concrete that contains no other cables. The brick or stone compartment for each phase of each switch should be so substantial that the contacts of that phase may arc to destruction without injury to the contacts of another phase. Bus-bars, like switches, should be removed from the operating switchboard, because an arc between them might destroy other apparatus thereon, and even the board itself. It is not enough to remove bus-bars from the switchboard where very high voltages are to be controlled, but each bar should be located in a separate brick compartment so that an arc cannot be started by accidental contact between two or more of the bars. It is convenient to have the brick and stone compartments for bus-bars built horizontally one above the other. The top and bottom of each compartment may conveniently be formed of stone slabs with brick piers on one side and a continuous brick wall on the other to hold the stone slabs in position. Connections to the bus-bars should pass through the continuous brick wall that forms what may be termed the back of the compartments. To close the openings between the brick piers at the front of the compartments movable slabs of stone may be used. Feeders passing away from the bus-bars, like dynamo cables running to these bars, should not be grouped close together in a single compartment, but each cable or circuit should be laid in a separate fireproof conduit to the point where it passes out of the station.

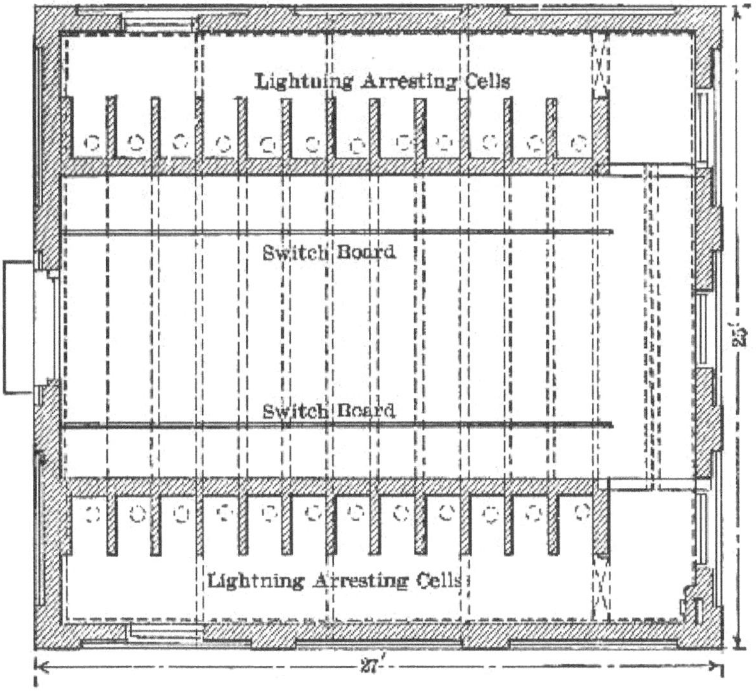

FIG. 60.—Ground Floor of Saratoga Switch-house.

The folly of grouping a large number of feeders that transmit great powers together in a single combustible compartment was well illustrated by the accident that destroyed the cables that connected the first Niagara power-station with the transformer-house on January 29th, 1903. On the evening of that day lightning short-circuited one of the cables in the short bridge that connects No. 1 station with the transformer-house, and all the cables in this bridge, supplying local consumers as well as railways and lighting in Buffalo, were destroyed. This bridge contained probably more than thirty-six cables, as that number of new cables was put in position within twenty-four hours after the accident, and these cables, covered with inflammable insulation, were close together. The result was not only the loss of the cables, but also the damage to power users. If these cables had been located in separate fire-proof conduits, it is highly probable that only the one directly affected by lightning would have been destroyed.

The brick and stone compartments for bus-bars may be located in the basement underneath the switchboard, as at the Portsmouth station of the New Hampshire Traction Company, or at any other place in a station where they are sufficiently removed from the other apparatus. In power-house No. 2 at Niagara Falls a cable subway beneath the floor level runs the entire length, parallel with the row of generators (A. I. E. E., vol. xix., p. 537). In this subway, which is thirteen feet nine and three-quarter inches wide and ten feet six inches high, the two structures for bus-bar compartments are located. Each of these structures measures about 6.6 feet high and 1.8 feet wide, and contains four bus-bar compartments. In each compartment is a single bar, and the four bars form two sets for two-phase working. Above the bus-bar compartments and rising from the floor level are the oil switches. A space over the cable subway midway of its length and between the two groups of oil switches is occupied by the switchboard gallery which is raised to some elevation above the floor and carries eleven generator, twenty-two feeder, two interconnecting, and one exciter panels. In power-house No. 1 the bus-bars are located in a common space above the 5,000 horse-power open-air switches already mentioned, and each bar has an insulation of vulcanized rubber covered with braid and outside of this a wrapping of twine. Of course; an insulation of this sort would amount to nothing if by any accident an arc were started between the bars. Where each bus-bar is located in a separate fireproof compartment, as at Niagara power-house No. 2, the application of insulation directly to each bar is neither necessary nor desirable. Consequently the general practice where each bar has its own fireproof compartment is to construct the bars of bare copper rods.

With main switches for generators and feeders removed from the operating board and actuated by electric motors or magnets, the small switches at the board with which the operator is directly concerned must of course control these magnets or motors. The small switches at the operating board are called relay switches, and the current in the circuits opened and closed by these switches and used to operate the magnets or motors of the oil switches may be conveniently obtained from a storage battery or from one of the exciting dynamos.

Probably the best arrangement of the relay switches is in connection with dummy bus-bars on the face of the switchboard, so that the connections on the face of the board constitute at all times a diagram of the actual connections of the generator and feeder circuits. It is also desirable for quick and correct changes in the connections of the main apparatus that all the relay switches and instruments necessary for the control of any one generator or any one feeder be brought together on a single panel of the switchboard. If this plan is followed, the operator at any time will have before him on a single panel all of the switches and instruments involved in the connections then to be made, and the chance for mistakes is thus reduced to a minimum. The plan just outlined was that adopted at the Niagara power plant No. 2, where a separate panel is provided for each of eleven generators and each of twenty-two feeders. On each of the eleven generator panels there are two selector relay switches, one generator relay switch, and one relay generator field switch. On each of the twenty-two feeder panels there are two relay selector switches. The relay switches on the two interconnecting panels serve to make connections between the two groups of five and six generators respectively in power-house No. 2 and the ten generators of power-house No. 1. On each panel there are relay indicators to show whether the oil switches that carry the main current respond to the movements of their relay switches.

Where the electric generators operate at the maximum voltage of the system, as at Garvin's Falls and in the power-house of the Manhattan Elevated Railway, there may be said to be only one general plan of connections possible. That is, the generators must connect directly with the main bus-bars at the voltage of the system, and the feeders or transmission lines must also connect to these same bars. Of course there may be several sets of bus-bars for different circuits or classes of work, but this does not change the general plan of through connections from generators to lines. So, too, the arrangement of switches is subject to variations, as by placing two switches in series with each other in each dynamo or feeder cable, or by connecting a group of feeders through their several switches to a particular set of bus-bars and then supplying this set of bars from the generator bus-bars through a single switch.

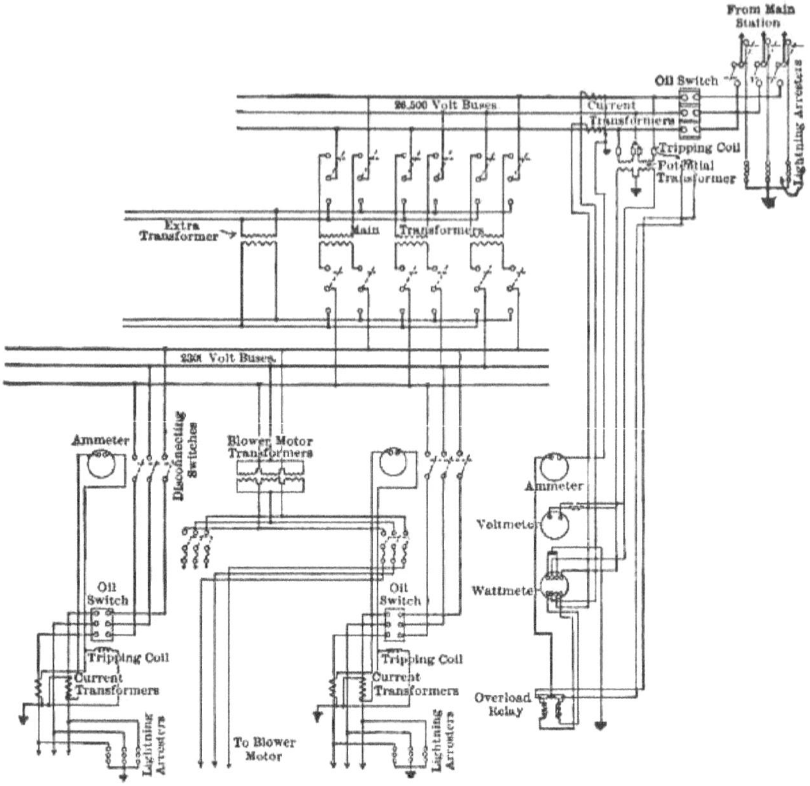

FIG. 61.—Switchboard Wiring, Glens Falls Sub-station on Spier Falls Line.

Where the voltage of transmission is obtained by the use of step-up transformers, the connections of these transformers may be such as to require nearly all switching to be done on either the high- or low-tension circuits. The more general practice was formerly to do all switching in the generator circuits and on the low-tension side of transformers, except in the connection and disconnection of transformers and transmission lines with the high-tension bus-bars, when not in operation. Where generators operate at the maximum voltage of the system only two main groups of switches are necessary, one group connecting generators to bus-bars, and the other group connecting bus-bars to the transmission lines. As soon as step-up transformers are introduced the number of switch groups must be increased to four if the usual method of connection is followed, and there must be both a high voltage and a low voltage set of bus-bars. That is, one set of switches must connect generators with low-tension bus-bars, another group must connect low-tension bars with the primary coils of transformers, a third group joins the secondary coils of transformers with the high-tension bars, and the fourth group of switches joins the transmission lines to the high-tension bus-bars. Switches connecting the secondary coils of step-up transformers to the high-tension bus-bars, and also the transmission lines to these same bars, have often been of the simple open-air type with short knife-blade construction. These switches have been used to disconnect the secondary coils of transformers and also the transmission lines from the high-tension bus-bars when no current was flowing, and switches of the simple knife-blade construction with short breaks could of course be used for no other purpose. With switches of this sort on the high-tension side of apparatus the practice is to do all switching of line circuits on the low-tension side.

It is possible to avoid some of this multiplication of switches if each generator with its transformers is treated for switching purposes as a unit and the switching for this unit is done on the secondary or high-voltage side of the step-up transformers. The adoption of this plan, of course, implies the use of switches that are competent to break the secondary circuit of any group of transformers under overload conditions and at the maximum voltage of the system, but oil switches as now made are competent to meet this requirement. When all switching of live circuits is confined to those of high voltage there is also the incidental advantage that heavy contact parts carrying very large currents are avoided in the operating switches. Where each generator is connected directly to its own group of transformers the secondary coils of these transformers will pass through oil switches to high-tension bus-bars, and the use of low-tension bus-bars may be avoided. From these high-tension bus-bars the transmission lines will pass through oil switches, so that on this plan there are only two sets of oil switches, namely, those connecting the secondary coils of transformers to the high-tension bus-bars, and those connecting the transmission lines to the same bars. Each group of two or three transformers, according as two or three are used with each generator, should be connected to its generator through short-break, open-air knife switches for convenience in disconnecting and changing transformers that are not in operation, but these switches are not intended or required to open the circuit of the generators and primary coils when in operation.

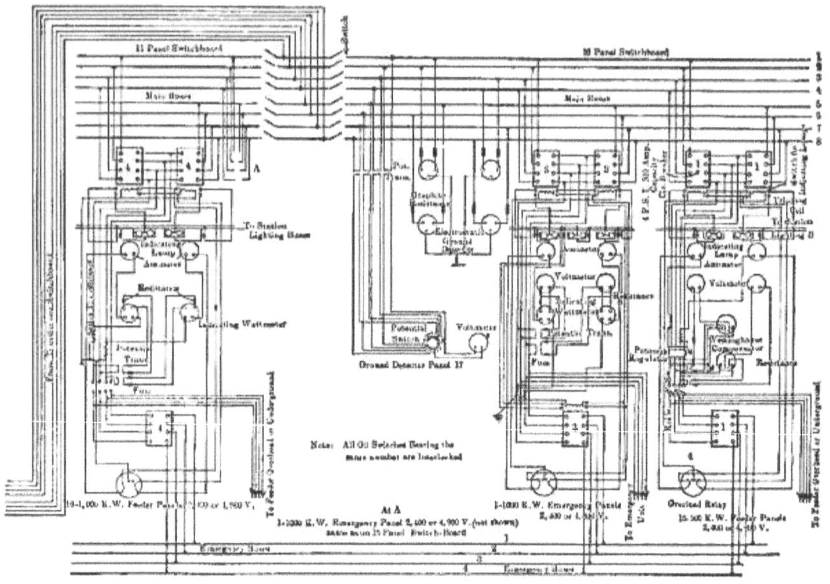

FIG. 62.—Distributing Switchboard, Central Sub-station, Montreal.

A plan similar to that just outlined was followed at the station of the Independent Electric Light and Power Company, San Francisco, where each of the 550-volt generators is ordinarily connected directly to the primary coils of two transformers that change the current from two-phase to three-phase and then deliver it through oil switches to the high-tension bus-bars at 11,000 volts. To these bus-bars the 11,000-volt feeders for five sub-stations are connected through switches. At this station there is a set of 550-volt bus-bars to which any of the generators may be connected, but to which no generator is connected in ordinary operation. The generators alone have switches connecting with these bars. When it is desirable to operate any particular generator on some pair of transformers other than its own, that generator is disconnected from its own transformers and connected to the 550-volt bus-bars. The generator whose transformers are to be operated by the generator before mentioned next has its switch connected to the 550-volt bus-bars, while the brushes of the contact rings of the former generator are raised. As the leads from each generator to its two switches are permanently joined, the switching operations just named connect the transformers of one generator with the other generator that has its switch closed on the 550-volt bars.

FIG. 63.—Switchboard at Chambly Power-station.

Where it is desired that a single reserve transformer may be readily substituted for any one of a number of transformers in regular use, the connections to each of these latter transformers may be provided with double-pole double-throw knife switches on both the primary and secondary sides, so that when these switches are thrown one way at any transformer in regular use the reserve transformer will be connected in its place.

Fuses and automatic circuit-breakers alike are intended to break connections without the intervention of human agency under certain predetermined conditions. In the fuse the heat generated by a certain current is sufficient to melt or vaporize a short length of special conductor. In the circuit-breaker a certain current gives a magnet or motor sufficient strength to overcome the pressure of a spring, and contact pieces through which the current is passing are pulled apart. The primary object of both the fuse and the circuit-breaker is thus to open connections and stop the flow of energy when more than a certain current passes. When any current passes through a circuit in the reverse of its regular direction the circuit-breaker can be arranged to break the connections, though the fuse cannot. A fuse must carry the current at which it is designed to melt during some seconds before enough heat is developed to destroy it, and the exact number of seconds for any particular case is made a little uncertain by the possibility of loose connections at the fuse tips which develop additional heat and also by the heat-conducting power of its connecting terminals. A circuit-breaker may be set so as to open its connections in one or more seconds after a certain current begins to flow. When connections are broken by a fuse the molten or vaporized metal forms a path that an arc may easily follow. A circuit-breaker with its contacts under oil offers a much smaller opportunity than a fuse for the maintenance of an arc. These qualities of fuses and circuit-breakers form the basis of their general availability and comparative advantages in transmission circuits.

Much variation exists in practice as to the use of fuses and circuit-breakers on transmission circuits. One view often followed is that fuses and circuit-breakers should be entirely omitted from the generator and transmission lines. The argument in favor of this practice is that temporary short circuits due to birds that fly against the lines or to sticks and loose wires that are thrown onto them will interrupt all or a large part of the transmission service if fuses or circuit-breakers that operate instantly are employed. On the other hand, it may be said that if fuses and circuit-breakers are omitted from the generator and transmission circuits a lasting short circuit will make it necessary to shut down an entire plant in some cases until it can be removed. Electric transmission at high voltages became important before magnetic circuit-breakers competent to open overloaded circuits at such voltages were developed. Consequently the early question was whether a transmission line and the generators that fed it should be provided with fuses or be solidly connected from generators to the distribution circuits of sub-stations. The original tendency was strong to use fuses in accord with the practice at low voltages. The great importance of continuous service from transmission systems and the many interruptions caused by temporary short circuits where fuses were used led to their abandonment in some cases. An example of this sort may be seen at the first Niagara station. In 1893, when this station was equipped, no magnetic circuit-breaker was available for circuits of either 11,000 or 2,200 volts, carrying currents of several thousand horse-power, and fuses were employed in lines at both these pressures (A. I. E. E., vol. xviii., pp. 495, 497). The fuses adopted in this case were the same for both the 2,200 and the 11,000-volt lines and were of the explosive type. Each complete fuse consisted of two lignum-vitæ blocks that were hinged together at one end and were secured when closed at the other. In these blocks three parallel grooves for fuses were cut and in each groove a strip of aluminum was laid and connected to suitable terminals at each end. Vents were provided for the grooves in which the aluminum strips were placed so that the expanding gas when a fuse was blown would escape. When these fuse blocks were new and the blocks of lignum vitæ made tight joints the metallic vapor produced when a fuse was blown was forced out at the vents and the connections of the line were thus broken. After a time, however, when the joints between the blocks were no longer tight because of shrinkage, the expanding gas of the fuse would reach the terminals and an arc would continue after the fuse had blown. These aluminum fuses, which were adopted about 1893, were abandoned at the Niagara plant in 1898. Since this later date the 2,200-volt feeders from the No. 1 power-house to the local consumers have had no fuses at the power-house, nor have circuit-breakers been installed there in the place of the fuses that were removed. At the large manufacturing plants supplied through these local Niagara feeders, the feeders formerly terminated in fuses, but these have since been displaced by circuit-breakers. In the second Niagara power-station, completed in 1902, the local 2,200-volt feeders are provided with circuit-breakers, but no fuses. Between the generators and bus-bars of the first Niagara plant the circuits were provided with neither fuses nor automatic circuit-breakers, and this practice continues there to the present time.

Besides the aluminum fuses in the 11,000-volt transmission line at the first Niagara station, there were lead fuses in the 2,200-volt primary circuits of the step-up transformers that supplied these lines. At the other end of these lines, in the Buffalo sub-station, another set of aluminum fuses was inserted before connection was made with the step-down transformers. Between the secondary coils of these transformers and the 550-volt converters there were no fuses, but these converters were connected to the railway bus-bars through direct current circuit-breakers. These lead fuses, which contained much more metal than those of aluminum, when blown set up arcs that lasted until power was cut off by opening a switch, and usually destroyed their terminals. An effort was made to so adjust the sizes of the fuses in this transmission system that in case of a short circuit in distribution lines at Buffalo only the fuses in the sub-station would be blown, leaving those at Niagara entire. This plan did not prove effective, however, and a severe overload on the distribution lines in Buffalo would blow out fuses clear back to the generator bus-bars at the Niagara station.

In order to accomplish the opening of overloaded circuits with greater certainty, to delay such opening where the overload might be of only a momentary nature, and to confine the open circuit to the lines where the overload existed, automatic circuit-breakers were substituted for the fuses named in the Niagara and Buffalo transmission system. This system was also changed from 11,000 to 22,000 volts on the transmission lines, thus rendering the requirements as to circuit-opening devices more severe. These circuit-breakers were fitted with time-limit attachments so that any breaker could be set to open at the end of any number of seconds after the current flowing through it reached a certain amount. A circuit-breaker with such a time-limit attachment will not open until the time for which it is set, after the amperes flowing through it reach a certain figure, has elapsed, no matter how great the current may be. Moreover, if the overload is removed from a line before the number of seconds for which its time-limit circuit-breaker is set have elapsed, the circuit-breaker resets itself automatically and does not open the connections. If a circuit-breaker is set to open a line after an interval of say three seconds from the time when its current reaches the limit, the line will not be opened by a mere momentary overload such as would blow out a fuse. By setting the time-limit relays of circuit-breakers in transmission lines to actuate the opening mechanism after three seconds from the time that an overload comes on, and then leaving the breakers on distribution lines to operate without a time-limit, it seems that the opening of breakers on the distribution lines should free the system from an overload there before the breakers on the transmission lines have time to act. Such a result is very desirable in order that the entire service of a transmission system may not be interrupted every time there is a fault or short circuit on one of its distribution lines. This plan was followed in the Niagara and Buffalo system. In the 22,000-volt lines at the Niagara station the time relays were set to actuate the breakers after three seconds, at the terminal house in Buffalo, where the transformers step down from 22,000 to 11,000 volts, the circuit-breakers in the 11,000 volt lines to sub-stations had their relays set to open in one second. Finally the circuit-breakers in the distribution lines from the several sub-stations were left to operate without any time limit. By these means it was expected that a short circuit in one of the distribution circuits from a sub-station would not cause the connections of the underground cable between that sub-station and the terminal house to be broken, because of the instant action of the circuit-breaker at the sub-station. Furthermore, it was expected that a short circuit in one of the underground cables between the terminal house and a sub-station would be disconnected from the transmission line at that house and would not cause the circuit-breakers at the Niagara station to operate. It is reported that the foregoing arrangement of circuit-breakers with time relays failed of its object because the breakers did not clear their circuits quick enough and that the time-limit attachments on the 22,000 and 11,000 volt lines are no longer in use (A. I. E. E., vol. xviii., p. 500). As the circuits under consideration convey thousands of horse-power at 11,000 and 22,000 volts it may be that time-limit devices with circuit-breakers would give good results under less exacting conditions. Time-limit relays are perhaps an important aid toward reliable operation of transmission systems, but they are subject to the objection that no matter how great the overload they will not open the circuit until the time for which they are set has run. In the case of a short circuit the time-limit relay may lead to a prolonged drop in voltage throughout the system, which is very undesirable for the lighting service and also allows all synchronous apparatus to fall out of step. With a mere momentary drop in voltage the inertia of the rotating parts of synchronous apparatus will keep them in step. For these reasons it is desirable to have circuit-breakers that will act immediately to open a line on which there is a short circuit or very great overload, but will open the line only after an interval of one or more seconds when the overload is not of a very extreme nature. This action on the part of circuit-breakers at the second Niagara power-station was obtained by the attachment of a dash-pot to the tripping plunger of each circuit-breaker (A. I. E. E., vol. xviii., p. 543). With moderate overloads of a very temporary nature this dash-pot so retards the action of a tripping plunger that the circuit-breaker does not open. When a short circuit or great overload comes onto a line the pull on the tripping plunger or the circuit-breaker on that line is so great that the resistance of the dash-pot to the movement is overcome at once and the line is disconnected from the remainder of the system.

The fact that a circuit-breaker may be designed to open the line which it connects, whenever the direction from which the flow of energy takes place is reversed, is taken advantage of at some sub-stations to guard against a flow of energy from a sub-station back toward the generating station. By this means a flow of energy from a sub-station to a short circuit in one of the lines or cables connecting it with the generating plant is prevented.

CHAPTER XII.
REGULATION OF TRANSMITTED POWER.

Regulation of voltage at incandescent lamps is a serious problem in the distribution of electrically transmitted energy. Good regulation should not allow the pressure at incandescent lamps rated at 110 to 120 volts to vary more than one volt above or below the normal.

Electric motor service is much less exacting as to constancy of voltage, and the pressure at motor terminals may sometimes be varied as much as ten per cent without material objection on the part of users. A mixed service to these three classes of apparatus must often be provided where transmitted energy is used, and the limitations as to variations at incandescent lamps are thus the ones that must control the regulation of pressure.

Transmission systems may be broadly divided into those that have no sub-stations and must therefore do all regulation at the generating plant, and those that do have one or more sub-stations so that regulation of voltage may be carried out at both ends of the transmission line.

FIG. 64.—Arc-lighting Switchboard at Central Sub-station, Montreal.

As a rule, a sub-station with an operator in attendance is highly desirable between transmission and distribution lines, and this is the plan generally followed at important centres of electrical supply, even though the transmission is a short one. One example of this sort may be noted at Springfield, Mass., where energy for electrical supply is transmitted from two water-power plants on the Chicopee River only about four and a half and six miles, respectively, from the sub-station in the business centre of the city. The voltage of transmission for two-phase current in this case is 6,000, and is reduced to about 2,400 volts at the sub-station for the general distribution of light and power. A similar instance may be seen at Concord, N. H., where electrical energy at both 2,500 and 10,000 volts is delivered to a sub-station in the business section from a water-power plant at Sewall's Falls, on the Merrimac River, four and one-half miles distant. From this sub-station the current is distributed at about 2,500 volts for the supply of lamps and motors. A sub-station was found desirable at Concord for purposes of regulation before the voltage of transmission was raised above that of distribution. Subsequently, when the load increased, the voltage of 10,000 was adopted on a part of the transmission circuit in order to avoid an increase in the size of their conductors.

FIG. 65.—Area of Electrical Distribution at Montreal.

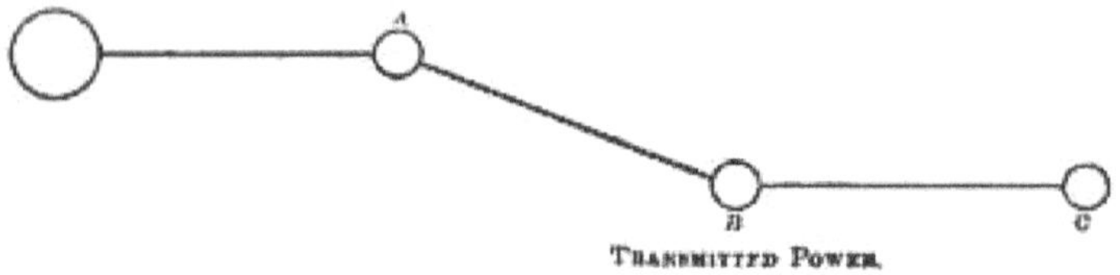

FIG. 66.

FIG. 67.

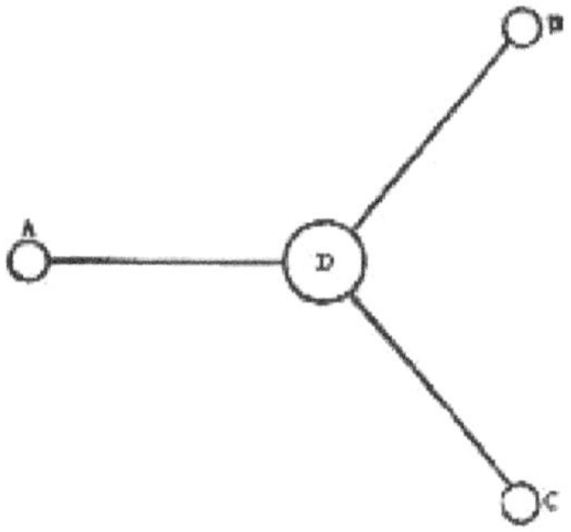

FIG. 68.

In some instances, however, transmission and distribution lines are joined without the intervention of a sub-station, where regulation of voltage can be accomplished, though this practice has little to recommend it aside from the savings in first cost of installation and subsequent cost of operation. These savings are more apparent than real if fairly constant pressure is to be maintained at the lamps, because what is gained by the omission of sub-stations will be offset, in part at least, by additional outlays on the lines if good regulation is to be maintained. This fact may be illustrated by reference to Figs. 66, 67, and 68, in each of which *D* represents a generating station and *A*, *B*, and *C* towns or cities where energy from the station is to be distributed. In the case of each figure it is assumed that the distance between the generating station and each of the cities or towns is such that distributing lines with a loss of, say, not more than two per cent in voltage at full load cannot be provided between the generating station and each city or town because of the cost of conductors. This being so, one or more centres of distribution must be located in each town, and the transmission lines must join the distribution lines at these centres either on poles or in sub-stations.

If several of these towns are in the same general direction from the generating plant so as to be reached by the same transmission line, as *A*, *B*, and *C* in Fig. 66, this one line will be all that is necessary with a sub-station in each town. Where sub-stations are not employed a separate transmission circuit must be provided between the generating plant and each town for reasons that will appear presently. The percentage of voltage variation in a transmission line under changing loads will be frequently from five to ten, and is thus far beyond the allowable variations at incandescent lamps. To give good lighting service the centre of distribution, where the transmission line joins the distribution circuits, must be maintained at very nearly constant voltage if no sub-station is located there. Regulation at a generating station will compensate for the changing loss of pressure in a line under varying loads so as to maintain a nearly constant voltage at any one point thereon. No plan of station regulation, however, can maintain constant voltages at several points on the same transmission line when there is a varying load at each. The result is that even though the several towns served are in the same general direction from the generating station, as in Fig. 67, yet each town should have its separate transmission line where no sub-stations in the towns are provided. In the case illustrated by Fig. 68, where the towns served are in very different directions from the generating station, there should be a separate transmission line to each, regardless of whether there is a sub-station or only a centre of distribution there.

Even in the case illustrated by Fig. 68, as in each of the others, there is a large saving effected in the cost of distribution lines by the employment of a sub-station at the point where these lines join the transmission circuit, provided that the variation of pressure at lamp terminals is to be kept within one volt either way from the standard. With the variations of loads the loss of pressure in the distribution lines will range from zero to its maximum amount and the connected lamps will be subjected to the change of voltage represented by this total loss, unless the distribution start from a sub-station where the loss in distribution lines can be compensated for by regulation. To give good service the distribution lines should be limited to a loss of one per cent at full load if there is no sub-station where they join transmission lines. With opportunity for regulation at a sub-station the maximum loss in distribution lines may easily be doubled, thus reducing their weight by one-half in comparison with that required where there is no sub-station.

Another advantage of connecting transmission and distribution lines in a sub-station, where regulation of voltage can be had, lies in the fact that it is practically impossible to maintain an absolutely constant pressure miles from a generating plant at the end of a transmission line that is carrying a mixed and varying load. A result is that without the intervention of regulation at a sub-station it is almost impossible to give good lighting service over a long transmission line. Furthermore, the labor of regulation at a generating station is much increased where there are no sub-stations, because it must be much more frequent and accurate. The absence of sub-stations from a transmission system thus implies more transmission circuits, heavier distribution circuits, more labor at the generating plant, and a poor quality of lighting service.

Where stationary motors form the great bulk of the load on a transmission system, and good lighting service is of small importance, it may be well to omit sub-stations at some centres of distribution. This is a condition that sometimes exists in the Rocky Mountain region where the main consumers of power along a transmission line may be mines or works for the reduction of ores. An example of this sort exists in the system of the Telluride Power Transmission Company, in Utah, which extends from Provo Cañon, on the river of the same name, entirely around Utah Lake by way of Mercur, Eureka, and Provo, and back to the power-house in Provo Cañon, a continuous circuit of 105 miles.

The transmission voltage on this line is 40,000, and at intervals where there are distributing points the voltage is reduced to about 5,000 by transformers on poles, and without the aid of regulation at sub-stations in some cases. The power thus transmitted is largely used in mines and smelters for the operation of motors, but also for some commercial lighting.

Regulation at generating stations of the voltage on transmission lines may be accomplished by the same methods whether there are sub-stations at centres of distribution or not. In any such regulation the aim is to maintain a certain voltage at some particular point on the transmission line, usually its end, where the distribution circuits are connected. If more than one point of distribution exists on the same transmission line, the regulation at the generating plant must be designed to maintain the desired pressure at only one of these points, leaving regulation at the others to be accomplished by local means. One method of regulation consists in the overcompounding of each generator so that the voltage at its terminals will rise at a certain rate as its load increases. If a generator and transmission line are so designed that the rise of voltage at the generator terminals just corresponds with the loss of voltage on the line when the output of that generator alone passes over it to some particular point, then the pressure at that point may be held nearly constant for all loads if no energy is drawn from the line elsewhere. These several conditions necessary to make regulation by the compounding of generators effective can seldom be met in practice. If a varying number of generators must work on the same transmission line, or if varying loads must be supplied at different points along the line, no compound winding of generators will suffice to maintain a constant voltage at any point on the line that is distant from the power-station. For these reasons the compound winding of generators is of minor importance so far as the regulation of voltage on transmission lines is concerned, and on large alternators is not generally attempted. An example may be noted on the 3,750-kilowatt generators at Niagara Falls, where the single magnet winding receives current from the exciters only.

A much more effective and generally adopted method of regulation of voltage at the generating plants of transmission systems is based on the action of an attendant who varies the current in the magnet coils of each generator so as to raise or lower its voltage as desired. The regulation must be for some one point on the transmission line, and the attendant at the generating plant may know the voltage at that point either by means of a pair of pressure wires run back from that point to a voltmeter at the generating plant, by a meter that indicates the voltage at the point in question according to the current on the line, or by telephone connection with a sub-station at the point where the constant voltage is to be maintained. Pressure wires are a reliable means of indicating in the generating station the voltage at a point of distribution on the line, but the erection of these wires is quite an expense in a long transmission, and in such cases they are only occasionally used. Owing to inductive effects and to variable power-factors the amperes indicated on a line carrying alternating current are far from a certain guide as to the drop in voltage between the generating station and the distant point. In long transmissions, telephone communication between the generating plant and the sub-stations is the most general way in which necessary changes to maintain constant voltage at sub-stations are brought to the attention of the attendant in the generating plant. Few, if any, extensive transmission systems now operate without telephone connection between a generating plant and all of its sub-stations, or between a single sub-station and the several generating plants that may feed into it. Thus, the generating plant at Spier Falls, on the Hudson River, will be connected by telephone with sub-stations at Schenectady, Albany, Troy, and some half-dozen smaller places. On the other hand, the single sub-station in Manchester, N. H., that receives the energy from four water-power plants has a direct telephone line to each.

Where two or more transmission lines from the same power-station are operated from the same set of bus-bars the voltage at a distant point on each line cannot be held constant by changes of pressure on these bus-bars. One generator only may be connected to each transmission line and be regulated for the loss on that line, but this loses the advantages of multiple operation. Another plan is to connect a regulator in each transmission line before it goes from the generating plant. One type of regulator for this purpose consists of a transformer with its secondary coil divided into a number of sections and the ends of these sections brought out to a series of contact segments. The primary coil of this transformer may be supplied with current from the bus-bars and the secondary coil is then connected in series with the line to be regulated, so that the secondary voltage is added to or subtracted from that of the main circuit. A movable contact arm on the segments to which the sections of the secondary coil are connected makes it possible to vary the secondary voltage by changing the number of these sections in circuit. In another transformer used for regulating purposes the primary coil is connected to the bus-bars as before and the movable secondary coil is put in series with the line to be regulated. The regulation is accomplished in this case by changing the position of the secondary relative to that of the primary coil and thus raising or lowering the secondary voltage. Both of these regulators require hand adjustment, and the attendant may employ the telephone, pressure wires, or the compensating voltmeter above mentioned, to determine the voltage at the centre of distribution. The voltage indicated by this so-called "compensator" is that at the generating station minus a certain amount which varies with the current flowing in the line to be regulated. The voltmeter coil of the compensator is connected in series with the secondary coils of two transformers, which coils work against each other. One transformer has its secondary coil arranged to indicate the full station voltage, and the other secondary coil is actuated by a primary coil that carries the full current of the regulated line. By a series of contacts the effect of this last-named coil can be varied to correspond with the number of volts that are to be lost at full load between the generating station and the point on the transmission line at which the voltage is to be held constant. If there is no inductive drop on the transmission line, or if this drop is of known and constant amount, the compensator may give the actual voltage at the point for which the regulation is designed.

Automatic regulators are used in some generating stations to maintain a constant voltage either at the generating terminals or at some distant distributing point on a line operated by a single generator. These regulators may operate rheostats that are in series with the magnet windings of the generators to be regulated, and raise or lower the generator voltage by varying the exciting current in these windings. These regulators are much more effective to maintain constant voltage at generating stations than at the distributing end of long transmission lines with variable power-factors. In spite of the compound winding of generators, of automatic regulators for the exciting currents in their magnet coils, and of regulating transformers in the transmission circuits, hand-adjustment of rheostats in series with the magnet coils of generators remains the most generally used at the generating stations of long transmission systems. Automatic regulators at the ends of transmission lines in sub-stations are now being introduced, and may prove very desirable.

FIG. 69.—Motor-generators in Shawinigan Sub-station at Montreal.

The more exacting and final work of regulation in transmission systems is usually done at the sub-stations. After a nearly constant voltage is delivered at the high-pressure coils of step-down transformers in a sub-station, there remains the varying losses in these transformers, in motor-generators or converters, in distribution lines and in service transformers, to be compensated for. In general, three or four sorts of loads must be provided for, namely, arc or incandescent lamps for street lighting on series circuits, usually of 4,000 to 10,000 volts. Arc and incandescent lamps on constant-pressure circuits of 2,000 to 2,500 volts for commercial lighting, direct-current stationary motors on constant-pressure circuits of about 500 volts, and alternating motors which may be served at either 2,500 or 500 volts according to their sizes and locations. To these loads may be added that of street-car motors of 500 volts, direct current. Both the stationary and the street-car motors, but more especially the latter, by their changes of load give rise to large and rapid fluctuations of voltage on the distribution lines to which they are connected. The problem of regulation with combined lamp and motor loads is not, therefore, so much to maintain a nearly constant voltage at the motors as to protect the lamps from the fluctuations of voltage which the motors set up.

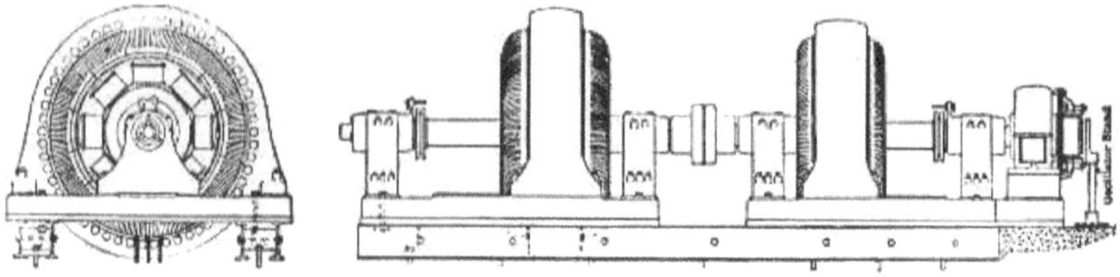

FIG. 70.—One of the 1,065-kilowatt Motor-generators in the Shawinigan Sub-station at Montreal.

For street-car motors using direct current at about 500 volts, the sub-station equipment includes either step-down transformers and converters or motor-generators with or without transformers. It is the practice in some cases where both lighting and street-railway service are drawn from the same transmission system to keep these two kinds of service entirely separate, devoting independent generators and transmission lines, as well as independent transformers and converters or motor-generators, to the street-car work. This is done in the transmission system centring at Manchester, N. H., in which each one of the four water-power plants, as well as the sub-station, has a double set of bus-bars on the switchboard; and from each water-power plant to the sub-station there are two transmission circuits. In operation, one set of generators, bus-bars, transmission circuits, and transformers supply converters or motor-generators for the street-car motors; and another set of generators, bus-bars, transmission circuits, and transformers are devoted to lighting and stationary motors in this system. Where street-car motors draw their energy from the same generators and transmission lines that supply commercial incandescent lamps, some means must be adopted to protect the lighting circuits from the fluctuations of voltage set up by the varying street-car loads. One way to accomplish this purpose is to operate the lighting circuits with generators driven by synchronous motors in the sub-stations. These generators may, of course, be of either direct or alternating type and of any desired voltage. The synchronous motors driving these generators take their current from the transmission line either with or without the intervention of step-down transformers. By this use of synchronous motors the lighting circuits escape fluctuations of voltage corresponding to those on the transmission line, because synchronous motors maintain constant speeds independently of the voltage of the circuits to which they are connected. This plan was followed at Buffalo, where the street-car system and the lighting service are operated with energy from the Niagara Falls stations over the same transmission line. In one of the sub-stations at Buffalo, both 2,200-volt, two-phase alternators, and 150-volt continuous-current generators for lighting service, are driven by synchronous motors connected to the Niagara transmission line through transformers. At other sub-stations in Buffalo, the 500-volt continuous current for street-car motors is obtained from the same transmission system through transformers and converters. Another solution of the problem of voltage regulation where street-railway and commercial lighting service are to be drawn from the same transmission line is found in the operation of 500-volt continuous-current generators in the sub-stations by synchronous motors fed from the line either directly or through transformers. This plan has been adopted in the transmission system of the Boston Edison Company, which extends to a number of cities and towns within a radius of twenty-five miles. The sub-stations at Natick and Woburn in this system, where there are street-railway as well as lighting loads, contain 500-volt continuous-current generators driven by synchronous motors connected directly to the three-phase transmission lines. In a case like this the synchronous motors maintain their speed irrespective of the voltage on the line and thus tend to hold that voltage steady in spite of the variable losses due to fluctuating loads.

Stationary motors should not as a rule be operated from the same distribution lines that supply incandescent lamps, especially in sizes above one horse-power, and this is the better practice. Motor circuits of about 2,400 volts and two- or three-phase, alternating, or 500 volts, alternating or direct current, may be supplied at a sub-station either by transformers alone in the first case or by transformers and converters in the second. In either case no especial provision is usually necessary for the regulation of constant pressure on the motor circuits.

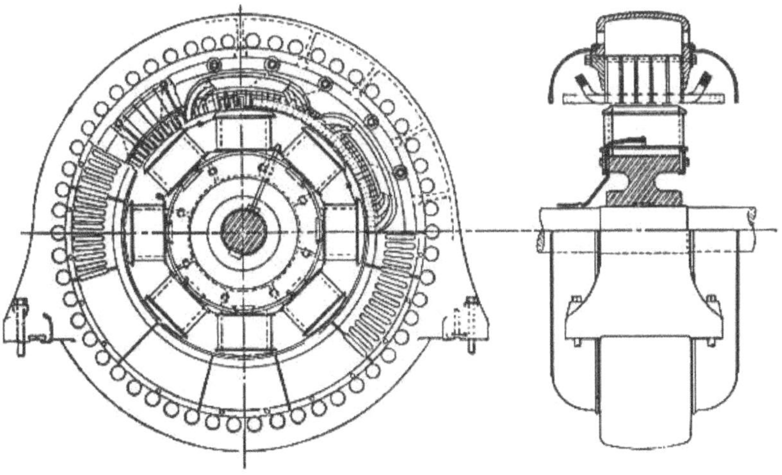

FIG. 71.—1,100-kilowatt, 2,300-volt, Three-phase, 30-cycle, Synchronous Motor at Sub-station of Shawinigan Line in Montreal.

In some transmission systems the distribution circuits for stationary motors are not fed by the same transmission lines that carry the lighting load, but draw their energy from lines that do no other work. This practice is certainly desirable, as it frees the lighting circuits from all fluctuations of voltage due to line losses with changing motor loads. Examples of this sort may be seen at Springfield, Mass., and Portland and Lewiston, Me., in each of which the load of stationary motors is operated over independent transmission as well as distribution lines.

In transmission systems series arc and incandescent lamps for street lighting are commonly operated either by direct-current arc dynamos or by constant-current transformers or constant-pressure transformers with automatic regulators at the sub-stations. The arc dynamos are driven by either induction or synchronous motors supplied directly from the transmission line or through transformers. As the arc dynamos regulate automatically for constant current no further regulation is required. If the series arc and incandescent lamps are to be supplied with alternating current, the constant-current transformer or the constant-current regulator come into use. This type of transformer and regulator alike depend for their regulating effect on the movement of a secondary coil on a transformer core in such a way that the current in this coil, which is in series with the lamps, is held nearly constant. Such constant-current transformers and regulators are usually supplied from the transmission line through regular constant-pressure transformers, and they hold their currents sufficiently constant for the purposes of their use.

The main problem of regulating thus comes back to the 250- or 2,200-volt, constant-pressure circuits for incandescent lighting, supplied from transmission lines through transformers or motor generators or both at the sub-station. For this regulation one of the most reliable instruments is the hand of a skilful attendant, guided by voltmeters connected with pressure wires from minor centres of distribution, and adjusting the regulating transformers above mentioned, or other regulating devices.

CHAPTER XIII.
GUARD WIRES AND LIGHTNING ARRESTERS.

Lightning in its various forms is the greatest danger to which transmission systems are exposed, and it attacks their most vulnerable point, that is, insulation. The lesser danger as to lightning is that it will puncture the line insulators and shatter or set fire to the poles. The greater danger is that the lightning discharge will pass along the transmission wires to stations and sub-stations and will there break down the insulation of generators, motors, or transformers. Damage by lightning may be prevented in either of two ways, that is, by shielding the transmission line so completely that no form of lightning charge or discharge can reach it, or by providing so easy a path from line conductors to earth that lightning reaching these conductors will follow the intended path instead of any other. In practice the shielding effect is sought by grounded guard wires, and the easy path for discharge takes the form of lightning arresters, but neither of these devices is entirely effective.

Aerial transmission lines are exposed to direct discharges of lightning, to electromagnetic charges due to lightning discharges near by, and to electrostatic charges that are brought about by contact with or induction from electrically charged bodies of air. It is evidently impracticable to provide a shield that will free overhead lines from all these influences. To cut off both electrostatic and electromagnetic induction from a wire and also to free it from a possible direct discharge of lightning, it seems that it would at least be necessary completely to incase the wire with a thick body of conducting material. This condition is approximated when an electric circuit is entirely beneath the surface of the ground, but would be hard to maintain with bare overhead wires. It seems, however, that grounded guard wires near to and parallel with long aerial circuits should tend to discharge any high electrostatic pressures existing in the surrounding air, and materially to reduce the probability that a direct discharge of lightning will choose the highly insulated circuits for its path to earth. Lightning arresters may conduct induced and direct lightning discharges to earth, without damage to transmission lines, so that both arresters and guard wires may logically be used in the same system.

Wide differences of opinion exist as to the general desirability of grounded guard wires on transmission lines, both because of their undoubted disadvantages and because the degree of protection that they afford is uncertain. It seems, however, that the defects of guard wires depend in large degree on the kind of wire used for the purpose, and the method of its erection. Galvanized iron wire with barbs every few inches has been more generally used for guard wires along transmission lines than any other sort.

Sometimes a single guard wire of this sort has been run on a pole line carrying transmission circuits, and the more common location of this single wire is on the tops of the poles. In other cases two guard wires have been used on the same pole line, one of these wires being located at each end of the highest cross-arm and outside of the power wires. Besides these guard wires at the ends of the top cross-arms of a pole line, a third wire has in some systems been added to the tops of the poles. These guard wires have sometimes been secured to the poles and cross-arms by iron staples driven over the wire and into the wood, and in other cases the guard wires are mounted on small glass insulators. Much variation in practice also exists as to the ground connections of guard wires, such connections being made at every pole in some systems, and much less frequently in some others.

With all these differences in the practical application of guard wires it is not strange that opinions as to their utility do not agree. Further reason for differences of opinion as to the practical value of guard wires exists in the fact that in some parts of the country the dangers from lightning are largely those of the static and inductive sort, that are most effectively provided for by lightning arresters, while in other parts of the country direct lightning strokes are the greatest menace to transmission systems. At the present time, knowledge of the laws governing the various manifestations of energy that are known under the general head of lightning is imperfect, and the most reliable rules for the use of guard wires along transmission lines are those derived from practical experience.

A case where a guard wire did not prove effective as a protection against lightning is that of the San Miguel Consolidated Gold Mining Company, of Telluride, Col., whose three transmission lines ran from the water-power plant to points from three to ten miles distant, as described in A. I. E. E., vol. xi., p. 337, and following pages. This transmission operated at 3,000 volts, single-phase, alternating, and the pole lines ran over the mountains at elevations of 8,800 to 12,000 feet above sea-level, passing across bare ridges and tracts of magnetic material. It was stated that the country over which the circuits ran is so dry and rocky that it was practically impossible to secure good ground connections along the line, and no mention was made of the way in which the ground wire was grounded, or of the number of its ground connections. Furthermore, it does not appear that there was more than one guard wire on each pole line. Under these circumstances, and with a certain make of lightning arresters in use at the station, lightning was a frequent cause of damage to the connected apparatus. The insulation of some of the machinery is described as honeycombed with perforations which led to continual leakage, grounds, and short-circuits, which seems to indicate that the damage was being done by static and inductive discharges rather than by direct lightning strokes, one of which would have disabled a machine at once.

The type of lightning arrester in use on this system was changed, and thorough ground connections were provided for the new arresters, after which the damage by lightning came to an end. It is not stated, however, that the guard wires were removed. This case has been referred to as one in which guard wires failed to give protection, but, as may be seen from the above facts, such a statement is hardly fair. In the first place, it does not appear that the single guard wire on each pole line was effectively grounded anywhere. Again, a large part of the damage to apparatus appears to have been the result of static or inductive discharges that could not in the nature of things have been prevented by a guard wire. Finally, as the guard wire was not removed after the new lightning arresters were erected, it is possible that this wire prevented some direct discharges over the transmission wires that would have been destructive.

On page 381 of the volume of A. I. E. E. above cited, it is stated that the frequency and violence of lightning discharges that entered a certain electric station on Staten Island were much less after guard wires had been erected along the connected circuits than they were before the guard wires were put up.

It is also stated on page 385 of the same volume that examination of statistics of a number of stations in this country and Europe had shown that in every case where an overhead guard wire had been erected over power circuits, or where these circuits ran for their entire distance beneath telegraph wires, lightning had given no trouble on the circuits so protected. Unfortunately, the speaker who made this statement did not tell us where the interesting statistics mentioned could be consulted.

On the first pole line erected for power transmission from Niagara Falls to Buffalo, two guard wires were strung at opposite ends of the top cross-arm on guard irons there located. This cross-arm also carried parts of two power circuits, and the nearest wires of these circuits were distant about thirteen inches from the guard irons. These guard wires were barbed, and grounded at every fifth pole, according to an account given in A. I. E. E., vol. xviii., at 514 and following pages. The character of the ground connections is not stated. Much trouble in the way of grounds and short circuits on the transmission lines was caused by these guard wires at times when they were broken by the weight of ice coatings and wind pressure. As a result of these troubles the guard wires were removed in 1898. Since that date it appears that the transmission lines between Niagara Falls and Buffalo have been without guard wires. Up to 1901, according to page 537 in the volume just cited, twenty per cent of the interruptions in operation at the Niagara plant were caused by lightning, and it seems probable that this record applies to the period after 1898, when the guard wires were removed. It is also stated that during a single storm the line was struck five times, and that five poles with their cross-arms were destroyed. If these direct lightning strokes occurred while there were no guard wires along the line, as seems to have been the fact, it is a fair question whether such wires well grounded would not have carried off the discharges without damage. In California, the country of long transmissions, the use of guard wires along the pole lines is quite general. Many of these lines run east and west across the State, and a single line may thus have elevations in its different parts all the way from that of tide-water up to several thousand feet above sea-level. Unless guard wires are strung with these lines there is much manifestation of induced or static electricity, according to an account at page 538, in vol. xviii., A. I. E. E., where it is said that in the absence of guard wires a person will be knocked off his feet every time he touches a transmission wire that is entirely disconnected from the source of power. It is also said that this static charge on idle power lines is sufficient, in time, to puncture the insulation of the connected apparatus. On the other hand, where the grounded and barbed guard wires are strung over the entire lengths of these long power lines, these lines may be handled with impunity when they are idle. Ground connections to the guard wire are said to be made at about every fourth pole, and to consist of a wire stapled down the face of the pole and joined to an iron plate beneath its butt. The barbed guard wire itself, of which each pole line appears to have but one, is regularly stapled to the tops of the poles.

At the reference just named it is related that on a certain transmission line running east and west across the State for a distance of forty-six miles, and protected by a guard wire, no trouble was experienced during a severe storm that swept north and south over the line. Meantime the damage on other lines in the same neighborhood, and presumably not protected by guard wires, was large.

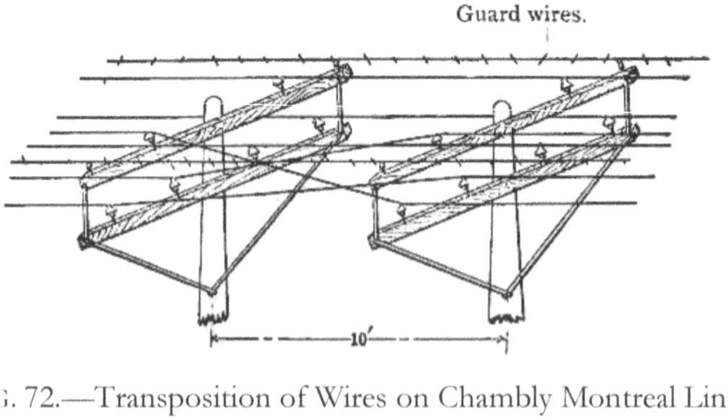

FIG. 72.—Transposition of Wires on Chambly Montreal Line.

Between the electric plant at Chambly, on the Richelieu River, and Montreal, Quebec, a distance of 16.6 miles, a transmission line of three circuits on two pole lines, with guard wires, was operated from some time in 1898 to December 1st, 1902, or somewhat more than four years. On the date last named the dam that maintained the head of water at the Chambly station gave way, and the plant was shut down during nearly a year for repairs. For as much as three years this line was operated at 12,000 volts, sixty-six cycles per second, two-phase. During the remainder of the period up to the failure of the dam the line was operated at 25,000 volts, sixty-three cycles, three-phase. In each transmission two pole lines were employed with two cross-arms per pole. One two-phase, four-wire circuit was carried on each of three of these cross-arms. At each end of the upper cross-arm on each pole, and at a distance of fifteen inches from the nearest power wire, a guard wire was mounted on a glass insulator. A third guard wire was mounted on a glass insulator at the top of each pole, and this third guard wire was about twenty inches from the nearest power wire. Each of these guard wires was made up of two No. 12 B. W. G. galvanized iron wires twisted together, with a four-point barb every five inches of length. Poles carrying these lines were ninety feet apart, and at each pole all three of the guard wires were connected by soldered joints to a ground wire that was stapled down the side of the pole, passed through an iron pipe eight feet long, and was then twisted several times about the butt of the pole before it was set in the ground. At three points along the line the conductors consisted of single-conductor underground or submarine cables that had an aggregate length of about twenty-five miles. No lightning arresters were employed at the points where the overhead transmission wires joined the underground cables.

These two-phase, 12,000-volt circuits were operated from some time in 1898 to some time in 1902, and during that time there was no damage done by lightning either at the Chambly plant, on the overhead line or the underground cable, or at the Montreal sub-station. This record is not due to lack of thunder-storms, for in the territory where the line is located these storms are frequent and severe. One very severe storm during the period in question resulted in serious damage on distribution lines at Chambly and Montreal, where the guard wires were not in use, but the transmission line and its connected apparatus escaped unharmed. The path of this storm was in the direction of the transmission line from Montreal to Chambly, and several trees were struck on the way. At the time of this storm and during an entire summer there were no lightning arresters in the power-house at Chambly.

In 1902, when the transmission line just considered was changed from two-phase to three-phase, and its voltage raised from 12,000 to 25,000, the method of protection by grounded, barbed guard wires, as above described, was retained. Two three-phase circuits were arranged on each of the two pole lines, with one wire of each circuit on an upper cross-arm and two wires of each circuit on a lower cross-arm, so that the nearest power wire on the upper cross-arm is thirty-two inches from the guard wire, and the nearest power wire on the lower cross-arm is about thirty inches from the guard wire at each end of the upper cross-arm. The guard wire at the tops of the poles is about thirty-three inches from each of the power wires on the upper cross-arm. In this three-phase line there is about 1,440 feet of three-conductor underground cable, and this cable lies between the end of the overhead line and the sub-station in Montreal. At the juncture of the overhead line and the cables there is a terminal house containing lightning arresters, and there are also arresters at the Chambly plant and the Montreal sub-station. No lightning arresters are connected to this line save those at the generating plant, the terminal house and the sub-station.

During that part of the year 1902 in which the new 25,000-volt line was in operation—that is, after the change and up to the time of the failure of the dam—this line and its connected apparatus were not damaged in any way by lightning, and the same is true for the period in which the line was idle pending repairs on the dam. The experience on this Montreal and Chambly transmission is probably among the best evidence to be found anywhere as to the degree of protection from lightning that may be had by the use of guard wires. In spite of cases like that just considered, where guard wires appear to have given a large degree of protection to transmission systems, many important transmissions are operated without them.

An example of this sort may be seen in the transmission line between the 10,000-horse-power plant at Electra, in the Sierra Nevada Mountains of California, and San Francisco, a distance of 154 miles, where it seems that no guard wires are in use. Another important transmission line that appears to get along without guard wires is that between the 10,000-horse-power plant at Cañon Ferry, on the Missouri River, and Butte, Mont., sixty-five miles away. On the transmission line between the power-station on Apple River, in Wisconsin, and the sub-station at St. Paul, Minn., about twenty-seven miles long, there are no guard wires for lightning protection. Further east, on the large, new transmission system that stretches from Spier Falls and Glens Falls on the north to Albany on the south, a distance in a direct line of forty miles, no guard wires are employed. On its way the transmission system just named touches Saratoga, Schenectady, Mechanicsville, Troy, and a number of smaller places, thus forming a network with several hundred miles of overhead wire. Examples of this sort might be multiplied, but those already named are sufficient to show that it is entirely practicable to operate long transmission systems without guard wires as a protection against lightning.

With these examples of transmission systems both with and without guard wires, the expediency of their use on any particular line should be determined by weighing their supposed advantages against their known disadvantages, under the existing conditions. It seems fairly certain from all the evidence at hand that if guard wires are to offer any large degree of protection to transmission systems such wires must be frequently and effectively grounded. There is certainly some reason to think that the failures of guard wires to protect transmission systems in some instances may have been due to the lack of numerous and effective ground connections. Such, for example, may have been the case above mentioned, at Telluride, Col. On the other hand, it seems reasonable to believe that the apparently high degree of protection afforded by the guard wires on the Chambly and Montreal line is due to the fact that these wires are connected through soldered joints at every pole with a ground wire that is wound about its base. The nearer the guard wires are located to the power wires on a line the greater is the danger that a guard wire will come into contact with a power wire by breaking or otherwise. It is probable that the protection given by a guard wire does not increase nearly as fast as the distance between it and a power wire is diminished. Even if one guard wire on a line is thought to be desirable, it does not follow that two or more such wires should be used, for the additional protection given by two or three guard wires beyond that given by one wire may be trifling, while the cost of erection and the danger of crosses with the power circuits increase directly with the number of guard wires. At one time it was thought very desirable to have barbs on guard wires, but now the better opinion seems to be that, as barbs tend to weaken the wire, they lead to breaks and cause more trouble than they are worth. The point where the barbs are located seems to rust more quickly than do other parts of the wire. In some cases barbed guard wires that have given trouble by breaking have been taken down and smooth wires put up instead. If a guard wire is well grounded at least as often as every other pole, its size may be determined largely on considerations of mechanical strength and lasting qualities. For ordinary spans a No. 4 B. & S. G. galvanized soft iron wire seems to be about right for guarding purposes. Iron seems to be the most desirable material for guard wires because it gives the required mechanical strength and sufficient conductivity at a less cost than copper, aluminum, or bronze, and is easier to handle and less liable to break than steel. It was formerly the practice to staple guard wires to the tops of poles or to the ends of cross-arms, but it was found that the wire was more apt to rust and break at the staple than elsewhere, and in the better class of work such wires are now mounted on small insulators. This practice, as stated above, was followed on the Montreal and Chambly line. In all cases the connection between the guard wire and each of its ground wires should be soldered, and the ground wire should have a large surface in contact with damp earth, either through a soldered joint with a ground plate, by winding a number of turns about the butt of the pole, or by other means.

It is thought by some telegraph engineers that the use of a separate ground wire running to the top of each pole is quite as effective as a protection against lightning as is a guard wire that runs to all of the poles and is frequently connected to the ground.

This practice is mentioned at page 26 of "Culley's Handbook of Practical Telegraphy." Such ground wires are free from most of the objections to the ordinary guard wires. It seems quite certain that a guard wire along an alternating-current line, and grounded at frequent intervals, must act as a secondary circuit of a transformer by reason of its ground connections, and thus absorb some energy from the power circuits. No experimental data are yet available, however, to show how large this loss may be in an ordinary case. It is fairly evident that there must be some electrostatic effects between the working conductors and a guard wire, but here again data are lacking as to the amount of any such effect. On most, if not all, transmission lines the guard wire or wires, if used at all, are placed either above or on a level with the highest power conductors. With one conductor of a three-phase circuit mounted on a pin set in the top of a pole, and the two remaining conductors on a two-pin cross-arm beneath, in the method most frequently adopted for transmission lines of very high voltage, it is obviously impracticable to put guard wires either above or on a level with the power circuits. In the latest transmissions there is a strong tendency to omit guard wires entirely and rely on lightning arresters for protection.

Lightning arresters are wrongfully named, for their true purpose is not to arrest or stop lightning, but to offer it so easy a path to the ground that it will not force its way through the insulation of the line or of machinery connected to the system. The requirements of a lightning arrester are in a degree conflicting, because the resistance of the path it offers must be so low as to allow discharges of atmospheric electricity to earth and so high as to prevent any flow of current between the transmission lines. In other words, the insulation of the line conductors must be maintained at a high standard in spite of the connection of lightning arresters between each conductor and the earth; but the resistance to the arrester must not be so high that lightning will pierce the insulation of the line or machinery at some other point. When a lightning discharge takes place through an arrester the resistance which the arrester offers to a flow of current is for the moment greatly reduced by the arcs which the lightning sets up in jumping the air-gaps of the arrester. Each wire of a transmission circuit must be connected alike to arresters, and the paths of low resistance through arcs in these arresters to the earth would obviously short-circuit the connected generators unless some construction were adopted to prevent this result. In some early types of lightning arresters magnetic or mechanical devices were resorted to in order to break arcs formed by the discharge of lightning.

The type of lightning arrester now in common use on transmission lines with alternating current includes a row of short, parallel, brass cylinders mounted on a porcelain block and with air-gaps of one-thirty-second to one-sixteenth of an inch between their parallel sides. The cylinder at one end of the row is connected to a line wire and the cylinder at the other end to the earth, when a 2,000 or 2,500-volt circuit is to be protected. For higher voltages a number of these single arresters are connected in series with each other and with the free ends of the series to a line wire and to the earth, respectively. Thus, for a 10,000-volt line, four or, better, five single arresters are connected in series to form a composite arrester for each line conductor. For any given line voltage the number of single arresters going to make up the composite arrester should be so chosen that the regular working voltage will not jump the series of air-gaps between the little brass cylinders, but yet so that any large rise of voltage will be sufficient to force sparks across these gaps. A variation of this practice by one large manufacturing company is to mount the group of single arresters on a marble board in series with each other and with an adjustable air-gap. This gap is intended to be so adjusted that any large increase of voltage on the lines will be relieved by a spark discharge. An arrester made up entirely of the brass cylinders and air-gaps has the disadvantage that an arc once started between all the cylinders by a lightning discharge so lowers the resistance between each line wire and the earth that the generating equipment is short-circuited and the arcs may not cease with the escape of atmospheric electricity. To avoid this difficulty it is the practice to connect a conductor of rather large ohmic resistance such as a rod of carborundum in series with the brass cylinders and air-gaps of lightning arresters. This resistance should be non-inductive so as not to offer a serious obstacle to lightning discharge, and its resistance should be great enough to prevent a flow of current from the generators that will be large enough to maintain the arcs started in the arrester by the lightning discharge. Accurate data are lacking as to the amount of this resistance that should be employed with arresters for any given voltage. As a rough, approximate rule it may be said that in some cases good results will be obtained with a resistance in ohms in series with a group of lightning arresters that represents one per cent of the numerical value of the line voltage. That is, for a 10,000-volt line the group of arresters for each wire may be connected to earth through a resistance of, say, 100 ohms, so that if the generator current follows the arc of a lightning discharge through the arresters it must pass through a fixed resistance of 200 ohms in going from one line wire to another. This rule is given merely as an illustration of the resistance that will work well in some cases, and should not be taken to have a general application. If the resistance connected in series with lightning arresters is high, the tendency is a little greater for lightning to go to earth at some point in the apparatus where the insulation is low. If only a small resistance is employed to connect lightning arresters with the earth, the danger is that arcs formed by lightning discharges will be followed and maintained by the dynamo currents. In one make of lightning arrester the row of little brass cylinders is connected at the ends to carbon rods which form a resistance for the purpose just mentioned. Two of these carbon rods are contained in each arrester for 2,000 or 2,500 volts, and the resistance of each rod may be anywhere from several score to several hundred ohms as desired. This form of arrester may be connected directly from line to earth without the intervention of any outside resistance, since the carbon rods may easily be given all the resistance that is desirable.

One of the most important features in the erection of a lightning arrester is its connection to earth. If this connection is poor it may render the arrester useless so far as protection from lightning is concerned. It need hardly be said that ground connections formed by driving long iron spikes into the walls of buildings or into dry earth are of slight value as far as protection from lightning is concerned. A good ground connection for lightning arresters may be formed with a copper or galvanized iron plate, which need not be over one-sixteenth of an inch thick, but should have an area of, say, ten to twenty square feet. This plate may be conveniently made up into the form of a cylinder and should have a number of half-inch holes punched in a row down one side into which one or more copper wires with an aggregate area equal to that of a No. 4 or No. 2 wire, B. & S. gauge, should be threaded and then soldered. This plate or cylinder should be placed deep enough in the ground to insure that the earth about it will be constantly moist, and the connected copper wire should extend to the lightning arresters. It is a good plan to surround this cylinder with a layer of coke or charcoal.

A good earth connection for lightning arresters may be made through large water-pipes, but to do this it is not enough simply to wrap the wire from the lightning arresters about the pipe. A suitable contact with such a pipe may be made by tapping one or two large bolts into it and then soldering the wires from lightning arresters into holes drilled in the heads of these bolts. A metal plate laid in the bed of a stream makes a good ground.

With some of the older types of lightning arresters it was the custom to insert a fuse between the line wire and the ground, but this practice defeats the purpose for which the arrester is erected because the fuse melts and leaves the arrester disconnected and the circuit unprotected with the first lightning discharge. The modern arresters for alternating-current circuits are made up of a series of metal cylinders and short air-gaps and are connected solidly without fuse between line and earth.

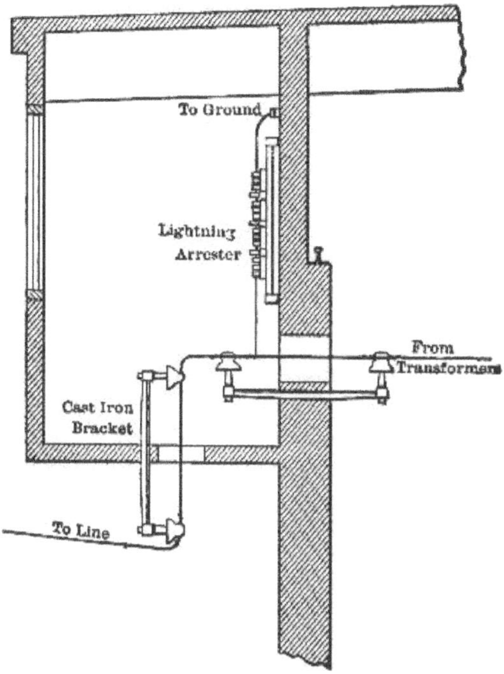

FIG. 73.—Entry of Lines at the Power-house on Neversink River.

It was once the practice to locate lightning arresters almost entirely in the stations, but this has been modified by experience and consideration of the fact that as the line acts as a collector of atmospheric electricity, paths for its escape should be provided along the line. Consideration fails to reveal any good reason why lightning that reaches a transmission line some miles from a station should be forced to travel to the station, where it may do great damage before it finds an easy path to earth. It is, therefore, present practice to connect lightning arresters to each wire at intervals along some lines as well as at stations and sub-stations. The main purpose of arresters is to offer so easy a path to earth that lightning discharges along the lines will not flow to points of low insulation in generators, transformers, or even the line itself. Practice is far from uniform as to the distance between lightning arresters on transmission lines, the distances varying from less than one to a large number of miles apart. In general the lines should be provided with lightning arresters at least where they run over hilltops and at any points where lightning strokes are unusually frequent. Where a long overhead line joins an underground cable arresters should always be connected, and the same is true as to transformers located on the transmission line. The multiplication of arresters along pole lines should be avoided as far as is consistent with suitable protection, because every bank of arresters may develop a permanent ground or short-circuit, unless frequently inspected and kept clean and in good condition.

Arresters, besides those connected along the lines, should be located either in or just outside of stations and sub-stations. If the buildings are of wood, the arresters had better be outside in weather-proof cases, but in brick or stone buildings the arresters may be properly located near an interior wall and well removed from all other station equipment. Transmission lines, on entering a station or sub-station, should pass to the arresters at once and before connecting with any of the operating machinery.

To increase the degree of protection afforded by lightning arresters choke-coils are frequently used with them. A choke-coil for this purpose usually consists of a flat coil of copper wire or strip containing twenty to thirty or more turns and mounted with terminals in a wooden frame. This coil is connected in series with the line wire between the point where the tap for the lightning arrester is made and the station apparatus. Lightning discharges are known to be of a highly oscillatory character, their frequency being much greater than that of the alternating currents developed in transmission systems. The self-induction of a lightning discharge in passing through one of these choke-coils is great, and the consequent tendency is to keep the discharge from passing through the choke-coil and into the station apparatus and thus to force the discharge to pass to earth through the lightning arrester. The alternating current employed in transmission has such a comparatively low frequency that its self-induction in a choke-coil is small. Increased protection against lightning is given by the connection of several groups of lightning arresters one after another on the same line wire at an electric station. This gives any lightning discharge that may come along the wire several paths to earth through the different groups of arresters, and a discharge that passes the first group will probably go to earth over the second or third group. In some cases a choke-coil is connected into a line wire between each two groups of lightning arresters as well as between the station apparatus and the group of arresters nearest thereto.

An electric transmission plant at Telluride, Col., where thunder-storms are very frequent and severe, was equipped with arresters and choke-coils of the type described, and the results were carefully noted (vol. xi., A. I. E. E., p. 346). A small house for arresters and choke-coils was built close to the generating station of this system and they were mounted therein on wooden frames. Four choke-coils were connected in series with each line wire, and between these choke-coils three lightning arresters were connected, while a fourth arrester was connected to the line before it reached any of the choke-coils. These arresters were watched during an entire lightning season to see which bank of arresters on each wire discharged the most lightning to earth.

It was found that, beginning on the side that the line came to the series of arresters, the first bank of arresters was traversed by only a few discharges of lightning, the second bank by more discharges than any other, the third bank by quite a large number of discharges, and the fourth bank seldom showed any sign of lightning discharge. Over the second bank of arresters the lightning discharges would often follow each other with great rapidity and loud noise. The obvious conclusion from these observations seems to be that three or four banks of lightning arresters connected in succession on a line at a station together with choke-coils form a much better protection from lightning than a single bank. At the plant in question, that of the San Miguel Consolidated Gold Mining Company, the entire lightning season after the erection of the arresters in question was passed without damage by lightning to any of the equipment. During the two lightning seasons previous to that just named the damage by lightning to the generating machinery at the plant had been frequent and extensive.

A good illustration of the high degree of security against lightning discharges that may be attained with lightning arresters and choke-coils exists at the Niagara Falls plants and the terminal house in Buffalo, where the step-up and step-down transformers have never been damaged by lightning though the transmission line has been struck repeatedly and poles and cross-arms shattered (vol. xviii., A. I. E. E., p. 527). This example bears out the general experience that lightning arresters, though not an absolute protection, afford a high degree of security to the apparatus at electric stations.

Lightning arresters are in some cases connected across high-voltage circuits from wire to wire so that the full line pressure tends to force a current across the air-gaps. The object of this practice is to guard against excessive voltages on the circuit such as might be due to resonance. In such a case, as in that where arresters are connected from line wire to earth as a protection against lightning, the number of air-gaps should be such that the normal line voltage will not force sparks across the air-gaps and thus start arcs between the cylinders.

The number and total length of air-gaps in a bank of arresters necessary to prevent the formation of arcs by the regular line voltage depends on a number of factors besides the amount of that voltage.

According to the report of the Committee on Standardization of the American Institute of Electrical Engineers, the sparking distances in air between opposed sharp needle points for various effective sinusoidal voltages are as follows (vol. xix., A. I. E. E., p. 1091):

Kilovolt Square Root of Mean Square.	Inches Sparking Distance.	
5	0	.225
10		.47
15		.725
20	1	.0
25	1	.3
30	1	.625
35	2	.0
40	2	.45
45	2	.95
50	3	.55
60	4	.65
70	5	.85
80	7	.1
90	8	.35
100	9	.6
110	10	.75
120	11	.85
130	12	.95
140	13	.95
150	15	.0

It may be noted at once from this table that the sparking distance between the needle points increases much faster than the voltage between them. Thus, 20,000 volts will jump an air-gap of only an inch between the points, but seven times this pressure, or 140,000 volts, will force a spark across an air-gap of 13.95 inches. Two cylinders or other blunt bodies show smaller sparking distances between them at a given voltage than do two needle points, but when a number of cylinders are placed in a row with short air-gaps between them the aggregate length of these gaps that will just prevent the passage of sparks at a given voltage may be materially greater or less than the sparking distance of that voltage between needle points. It has been found by experiment that the numbers one-thirty-second-inch spark-gaps between cylinders of non-arcing alloy necessary to prevent the passage of sparks with the voltages named and a sine wave of electromotive force are as follows (vol. xix., A. I. E. E., p. 1026):

Number of 1/32-Inch Air-Gaps in Series.	Normal Voltage Withheld.
5	6,800
10	10,000
15	12,500
20	14,500
25	16,400
30	18,200
35	19,300
40	20,500
45	21,700
50	22,600
55	23,900
60	25,000
65	26,000
70	27,000
75	28,000
80	29,000

According to these data, only ten air-gaps of one-thirty-second of an inch each and 0.3125 inch combined length are required between cylinders to prevent a discharge at 10,000 volts, though opposed needle points may be 0.47 inch apart when a spark is obtained with this voltage. On the other hand, eighty air-gaps of one-thirty-second of an inch each between cylinders of non-arcing metal, or a total gap of 2.5 inches, are necessary to prevent a discharge at 29,000 volts, though 30,000 volts can force a spark across a single gap of only 1.625 inches between opposed needle points.

Under the conditions that existed in the test just recorded the pressure at which the aggregate length of one-thirty-second of an inch air-gaps that just prevents a discharge equals the single sparking distance between needle points seems to be about 18,000 volts.

The object of dividing the total air-gap in a lightning arrester for lines that carry alternating current up into a number of short gaps is to prevent the continuance of an arc by the regular generator or line current after the arc has been started by a lightning discharge. As soon as an electric spark leaps through air from metal to metal, a path of low electrical resistance is formed by the intensely heated air and metallic vapor. If the arc thus formed is, say, two inches long it will cool a certain amount as the passing current grows small and drops to zero. If, however, this total arc of two inches is divided into sixty-four parts by pieces of metal, the process of cooling as the current decreases will go on much more rapidly than with the single arc of two inches because of the great conducting power of the pieces of metal. As an alternating current comes to zero twice in each period, the many short arcs formed in an arrester by a lightning discharge are so far cooled during the small values of the following line current that the resistance quickly rises to a point where the regular line voltage cannot continue to maintain them, if the arrester is properly designed for the system to which it is connected. In this way the many-gap arrester destroys the many small arcs started by lightning discharges that would continue and short-circuit the line if they were combined into a single long arc.

When an electric arc passes between certain metals like iron and copper a small bead is raised on their surfaces. If these metals were used for the cylinders of arresters the beads on their surface would quickly bridge the short air-gaps. Certain other metals, like zinc, bismuth, and antimony, are pitted by the passage of arcs between their surfaces. By suitable mixture of metals from these two classes, an alloy is obtained for the cylinders of lightning arresters that pits only slightly and is thus but little injured by lightning discharges. After long use and many discharges an arrester of the class here considered gradually loses its power to destroy electric arcs. This may be due to the burning out of the zinc and leaving a surface of copper on the cylinders.

Aside from the structure of an arrester and the normal voltage of the circuit to which it is connected, its power to destroy arcs set up by lightning discharges depends on the capacity of the connected generators to deliver current on a short-circuit through the gaps, and upon the inductance of the circuit. The greater the capacity of the generators connected to a system the more trying are the conditions under which arresters must break an arc because the current to be broken is greater. So, too, an increase of inductance in a circuit adds to the work of an arrester in breaking an arc.

An arc started by lightning discharge at that period of a voltage phase when it is at or near zero is easily destroyed by the arrester, but an arc started at the instant when the regular line voltage has its maximum value is much harder to break because of the greater amount of heat generated by the greater current sent through the arrester. For this reason the arcs at arresters will hold on longer in some cases than in others, according to the portion of the voltage phase in which they are started by the lightning discharge. Lightning discharges, of course, may occur at any phase of the line voltage, and for this reason a number of discharges must take place before it can be certain from observation that a particular arrester will always break the resulting arc. Between twenty-five and sixty cycles per second there is a small difference in favor of the latter in the power of a given arrester to break an arc, due probably to the fact that more heat in the arcs is developed per phase with the lower than with the higher frequency.

It will now be seen that while increase of the regular line voltage requires a lengthening of the aggregate air-gap in lightning arresters to prevent the formation of arcs by this voltage alone, the increase of generating capacity requires more subdivisions of the total air-gap in order that the arcs maintained by the larger currents may be cooled with sufficient rapidity. These two requirements are to some extent conflicting, because the subdivision of the total air-gaps renders it less effective to prevent discharges due to the normal line voltage, as has already been shown. The result is that the more an air-gap is subdivided in order to cool and destroy arcs that have been started by lightning, the longer must be the aggregate air-gap in order to prevent the development of arcs directly by the normal line voltage.

Furthermore, the practical limit of subdivision of the air-gap is soon reached because of the difficulty of keeping very short gaps clean and of nearly constant length. As a resistance in series with an arrester cuts down the generator current that can follow a lightning discharge, such a resistance also decreases the number of air-gaps necessary to give an arrester power to destroy arcs on a particular circuit.

The increase of resistance in series with a lightning arrester as well as the increase in the aggregate length of its air-gaps subjects the insulation of connected apparatus to greater strains at times of lightning discharge. On systems of large capacity the number and aggregate length of air-gaps in arresters necessary to destroy arcs must be greater than the number or length of these air-gaps necessary to prevent the development of arcs by the normal line voltage, unless a relatively large resistance is connected in series with each arrester. To reduce the strains produced on the insulation of line and connected apparatus under these conditions by lightning discharges, a resistance is connected in shunt with a part of the air-gaps in one make of lightning arrester. The net advantage claimed for this type of arrester is that a lower resistance may be used in series with all the air-gaps than would otherwise be necessary. One-half of the total number of air-gaps in this arrester are shunted by the shunt resistance and the series and shunt resistance are in series with each other. Only the series air-gaps or those that are not shunted must be jumped in the first instance by the lightning discharge, which thus passes to earth through these air-gaps and the shunt and series resistance in series. An arc is next started in the shunted air-gaps, and this arc is in turn destroyed because the shunt weakens the current in these gaps. This throws the entire current of the arc through the series air-gaps and the shunt and series resistance all in series with each other. As the shunt resistance is comparatively large, the current maintaining the arc in the series air-gaps is next so reduced that this arc is broken. Taking the claims of its makers just as they stand, the advantages of the shunted air-gaps are not very clear. The series air-gaps alone must evidently be such that the normal line voltage will not start an arc over them, and these same series gaps must be able to break the arcs of line current flowing through them and the shunt and series resistance all in series. Evidently the greatest strain on the insulation of the line and apparatus occurs at the instant when the lightning discharge takes place through the series gaps and the shunt and series resistances all in series with each other.

Why develop subsequent arcs in the shunted air-gaps? Why not throw the shunted air-gaps away and combine the shunt and series resistances?

CHAPTER XIV.

ELECTRICAL TRANSMISSION UNDER LAND AND WATER.

Energy transmitted over long distances must sometimes pass through conductors that are underground or beneath water. In some other cases it is a question of relative advantages merely, whether portions of a transmission line go under water or overhead. Where the transmitted energy must enter a sub-station in the heart of a large city, it not infrequently must go by way of underground conductors without regard to the voltage employed. In some cities the transmission lines may be carried overhead, provided that their voltage is within some moderate figure, but not otherwise. Here it becomes a question whether transmission lines at high voltage shall be carried underground, or whether transforming stations shall be established outside of the restricted area and then low-pressure lines brought into the business section overhead or underground, as desired. Where a transmission line must cross a steam railway track it may be required to be underground, whether the voltage is reduced or not. The distance across a body of water in the path of a transmission line may be so great that a span is impossible and a cable under the water therefore necessary. Such a cable may work at the regular line voltage, or a transformer station may be established on one side or on each side of the body of water. Even where it is possible to span a body of water with a transmission line, the cost of the span and of its supports may be so great that a submarine cable is more desirable. A moderate increase in the length of a transmission line in order to avoid the use of a submarine cable is almost always advisable, but where rivers are in the path of the line it is generally impossible to avoid crossing them either overhead or underneath. Thus, St. Paul could only be reached with the 25,000-volt line from the falls on Apple River by crossing the St. Croix River, one-half mile wide, on the way. In order to carry out the 40,000-volt transmission between Colgate and Oakland, the Carquinez Straits, which intervened with nearly a mile of clear water, were crossed. Sometimes, as in the former of the two cases just named, an existing bridge may be utilized to support a transmission line, but more frequently the choice lies between an overhead span from bank to bank of a river and a submarine cable between the same points.

The prime advantage of an overhead line at high voltage is its comparatively small first cost, which is only a fraction of that of an underground or submarine cable in the great majority of instances. At very high voltages, like 40,000 to 50,000 or more, the overhead line must also be given first place on the score of reliability, since the lasting qualities of underground and submarine cables at such pressures is as yet an unknown quantity. On the other hand, at voltages in which cable insulation has been shown by experience to be thoroughly effective, underground or submarine cables may be more reliable than overhead lines because of the greater freedom from mechanical disturbances which these cables enjoy.

In the business portion of many cities a transmission line must go underground, whether its voltage is high or low. Under these conditions, it may be desired either to transmit energy to a sub-station for distribution within the area where conductors must be underground, or to transmit energy from a generating station there located to outside points. If the transmitted energy is reduced in pressure before reaching such a sub-station, a transforming station must be provided, and this will allow the underground cables to operate at a moderate voltage. For such a case the advantages as to insulation at the lower voltage should be compared with the additional weight of conductors in the cable and the cost of the transforming apparatus and station. If the voltage at which current is delivered from the transforming station does not correspond with the required voltage of distribution at the sub-station, the necessary equipment of step-down transformers is doubled in capacity by lowering the voltage of the transmitted energy where it passes from the overhead line to the underground cables. Conditions of just this sort exist at Buffalo in connection with the delivery of energy from the power-stations at Niagara Falls. This transmission was first carried out at 11,000 volts, and a terminal station was located at the Buffalo city limits where the overhead lines joined underground cables that continued the transmission at the same voltage to several sub-stations in different parts of the city. Later the voltage of the overhead transmission line was raised to 22,000, and it not being thought advisable to subject the insulation of the underground cables to this higher pressure, transformers were installed at the terminal station to lower the line voltage to 11,000 for the underground cables. As the sub-stations in this case also have transformers, there are two kilowatts of capacity in step-down transformers for each kilowatt of delivery capacity at the sub-stations.

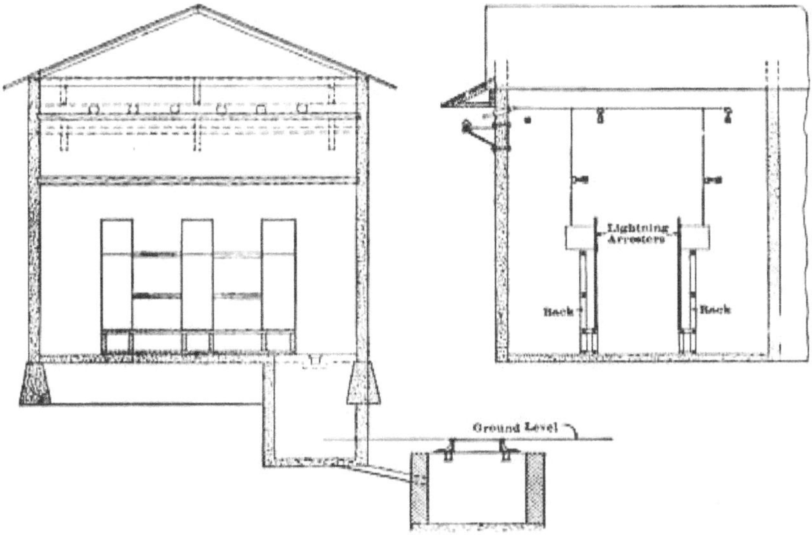

FIG. 74.—Cable Terminal House for the 25,000-volt Chambly Line at Montreal.

The saving effected in capacity of transformers and in the weight of cables by continuing the full transmission voltage right up to the sub-stations whence distribution takes place furnishes a strong motive to work underground cables at the pressure of the overhead transmission line of which they form a continuation. Thus, at Hartford, the 10,000-volt overhead lines that bring energy from water-power stations to the outskirts of the city connect directly in terminal houses there with underground cables that complete the transmission to the sub-station at the full line voltage. In Springfield, Mass., the overhead transmission lines from water-power stations connect directly with underground cables at a distance of nearly two miles from the sub-station, and these cables are thus subject to the full line pressure of 6,000 volts. The overhead line that brings energy at 25,000 volts from Apple River falls to St. Paul terminates about three miles from the sub-station there, and the transmission is completed by underground cables that carry current at the 25,000-volt pressure.

In these and similar cases the relative advantages of underground cables at the full voltage of transmission and of overhead lines at a much lower pressure, in the central portions of cities, must be compared. The overhead lines at moderate voltage will no doubt cost less in almost every case than underground cables of equal length and at the full transmission voltage.

As an offset to the lower cost of overhead city lines at moderate voltage, where they are permitted by local regulations, comes the increase in weight of conductors due to the low pressure on the overhead lines, and also the cost of additional transformer capacity, unless the lines that complete the transmission operate at the voltage of distribution. The 10,000-volt lines that transmit energy from Great Falls to Portland, Me., terminate in two transformer houses that are distant about 0.5 mile and 2.5 miles, respectively, from the sub-station there. In these transformer houses the voltage is reduced to 2,500, and the transmission is then continued at this pressure to the sub-station whence distribution takes place without further transformation.

Where a river or other body of water must be crossed by a transmission line, either of three plans may be followed. The overhead line may be continued as such across the water, either by a single span or by two or more spans supported by one or more piers built for that purpose in the water. The overhead line may connect directly with a submarine cable, this cable being thus exposed to the full voltage of the transmission. As a third expedient, a submarine cable may be laid and connected with step-down transformers on one bank and with step-up transformers on the other bank of the river or other body of water to be crossed. The overhead lines connecting with these transformers can obviously be operated at any desired voltage, and this is also true of the cable.

Even though the distance across a body of water is not so great that a transmission line can not be carried over it in a single span, the cost of such a span may be large. A case in point is that of the Colgate and Oakland line, where it crosses the Carquinez Straits by a span of 4,427 feet. These straits are about 3,200 feet wide where the transmission line crosses, and overhead lines were required to be not less than 200 feet above high water so as not to impede navigation. In order to gain in ground elevation and thus reduce the necessary height of towers, two points 4,427 feet apart on opposite sides of the straits were selected for their location. Under these circumstances two steel towers, one sixty-five feet and the other 225 feet high, were required to support the four steel cables that act as conductors across the straits. To take the strain of these four cables, each with a clear span nearly three times as great as that of the Brooklyn Bridge, eight anchors with housed strain insulators were constructed, four on the land side of each tower. On each of these anchors the strain is said to be 24,000 pounds. At each end of the cables making this span is a switch-house where either of the two three-phase transmission lines may be connected to any three of the four steel cables, thus leaving one cable free for repairs. It is not possible to state here the relative cost of these steel towers and cables in comparison with that of submarine cables for the same work, but at first glance the question appears to be an open one. The voltage of 40,000, at which this transmission is carried out, is probably higher than that on any submarine cable in use, but it is possible that a suitable cable can be operated at this voltage. Whatever the limitations of voltage as applied to submarine cables, it would, of course, have been practicable to use step-up and step-down transformers at the switch-houses and thus operate a submarine cable at any voltage desired.

In another case, on a transmission between Portsmouth and Dover, N. H., it was necessary to cross an arm of the sea on a line 4,811 feet long with a three-phase circuit operating at 13,500 volts. It was decided to avoid the use of either a great span or of raising and lowering transformers at this crossing, and to complete the line through a submarine cable operating at the full voltage of transmission. To this end a brick terminal house six by eight feet inside, and with an elevation of thirteen feet from the concrete floor to the tile roof, was erected on each bank of the bay at the point where the submarine cable came out of the water. The lead-covered cable pierced the foundation of each of these terminal houses at a point four feet below the floor level and rose thence on one wall to an elevation eleven feet above the floor to a point where connection was made with the ends of the overhead lines. From this connection on each of the three conductors a tap was carried to a switch and series of lightning arresters. A single lead-covered cable containing three conductors makes connection between these two terminal houses. At each end of this cable the lead sheath joins a terminal bell one foot long and 2.5 inches in outside diameter, increasing to four inches at the end where the three conductors pass out. This terminal bell is filled nearly to the flaring upper end with an insulating compound.

In the instance just named it is possible that the cost of the submarine cable was less than would have been the outlay for shore supports and a span nearly a mile long across this body of water.

Underground and submarine cables have been operated at voltages suitable for transmission during periods sufficiently long to demonstrate their general reliability. The Ferranti underground cables between Deptford and London have regularly carried current at 11,000 volts since a date prior to 1890. During about five years cables with an aggregate length of sixteen miles have transmitted power from St. Anthony's Falls to Minneapolis. At Buffalo, some thirty miles of rubber-insulated cables have been in use for underground work at 11,000 volts since 1897, and eighteen miles of paper-insulated cable since the first part of 1901. These examples are enough to show that transmission through underground cables at 11,000 to 12,000 volts is entirely practicable. At Reading, Pa., an underground cable one mile long has carried three-phase current at 16,000 volts for the Oley Valley Railway since some time in 1902. The cables in the transmission from Apple River to St. Paul, which carry three-phase current at 25,000 volts, have a total length of three miles, and have been in use since 1900. This voltage of 25,000 is probably the highest in regular use on any underground or submarine cable conveying energy for light or power. From the experience thus far gained there is much reason to think that the voltages applied to underground cables may be very materially increased before a prohibitive cost of insulation is reached.

On submarine cables the voltage of 13,000 in the Portsmouth and Dover transmission, above mentioned, is perhaps as great as any in use. It does not appear, however, that any material difference exists, as to the strain on its insulation at a given voltage, between a cable when laid in an underground conduit and when laid under water. In either case the entire stress of the voltage employed operates on the insulation between the several conductors in the cable and between each conductor and the metallic sheath. Underground conduits have little or no value as insulators of high voltages, because it is practically impossible to keep them water-tight and prevent absorption or condensation of moisture therein. For these reasons a cable that gives good results at 25,000 volts in an underground conduit should be available for use at an equal voltage under water. The standard structure of high-voltage cables for either underground or submarine work includes a continuous metallic sheath outside of each conductor or of each group of conductors that goes to make up a circuit. As most transmissions are now carried out with three-phase current, the three conductors corresponding to a three-phase circuit are usually contained in a single cable and covered by a single sheath. The cables used in transmission systems at Portsmouth, Buffalo, and St. Paul are of this type. If single-phase or two-phase current is transmitted, each cable should contain the two conductors that go to make up a circuit. In work with alternating currents the use of only one conductor per cable should be avoided because of the loss of energy that results from the currents induced in the metallic sheath of such a cable.

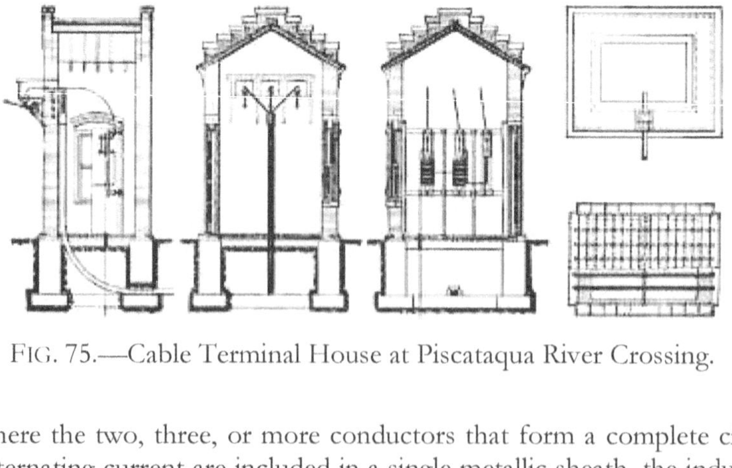

FIG. 75.—Cable Terminal House at Piscataqua River Crossing.

Where the two, three, or more conductors that form a complete circuit for alternating current are included in a single metallic sheath, the inductive effects of currents in the several conductors tend to neutralize each other and the waste of energy in the sheath is in large part avoided. To neutralize more completely the tendency to local currents in their metal sheath, the several insulated conductors of an alternating circuit are sometimes twisted together, after being separately insulated, before the sheath is put on. Distribution of power at Niagara Falls was at first carried out through single-conductor, lead-covered cables with two-phase current at 2,200 volts. One objection to this plan was the loss of energy by induced currents in the lead coverings of the cables. It was later decided to adopt three-phase distribution at 10,000 volts for points distant more than two miles from the power-station. Each three-phase circuit for this purpose was made up of three conductors separately insulated and then covered with a single lead sheath, so as to avoid losses through induced currents in the latter. Underground and submarine cables for operation at high voltages are generally covered with a continuous lead sheath and sometimes with a spiral layer of galvanized iron wire. For high-voltage work underground the lead covering is generally preferred without iron wire, but in submarine work coverings of both sorts are employed. The lead sheath of a cable being continuous completely protects the insulation from contact with gases or liquids. As ducts of either tile, wood, or iron form a good mechanical protection for a cable, the rather small strength of a lead sheath is not a serious objection in conduit work. Submarine cables, on the other hand, depend on their own outer coverings for mechanical protection, and may be exposed to forces that would rapidly cut through a lead sheath. Cables for operation under water should usually be covered, therefore, with a layer of galvanized iron wires outside of the lead sheath. These wires are laid closely about the cable in spiral form and are usually between 0.12 and 0.25 inch in diameter each, depending on the size of the cable and its location.

Underground conduits cannot be relied on to exclude moisture and acids of the soil from the cables which they contain, and either of these agents may lead to destructive results. If cables insulated with rubber, but without a protecting covering outside of it, are laid in underground conduits, the rubber is apt to be rapidly destroyed by fluids and gases that find their way into the conduit. If a plain lead-covered cable is employed the acids of the soil attack it, and if stray electric currents from an electric railway find the lead a convenient conductor it is rapidly eaten away where they flow out of it. To avoid both of these results the underground cable should have a lead sheath, and this sheath may be protected by an outside layer of hemp or jute treated with asphaltum.

Rubber, paper, and cotton are extensively used as insulation for underground and submarine cables, but the three are not usually employed together. As a rule, the insulation is applied separately to each conductor, and then an additional layer of insulation may be located about the group of conductors that go to make up the cable. Where rubber insulation is employed, a lead sheath may or may not be added, but where insulation depends on cotton or paper the outer covering of lead is absolutely necessary to keep out moisture. The radial thickness of insulation on each conductor and of that about the group of conductors in a cable should vary according to the voltage of operation.

Cables employed between the generating and sub-stations of the Manhattan Elevated Railway, to distribute three-phase current at 11,000 volts, are of the three-conductor type, rubber insulated, lead covered, and laid in tile conduits. Each cable contains three No. 000 stranded conductors, and each conductor has its own insulation of rubber. Jute is laid on to give the group of conductors an outer circular form, and outside of the group a layer of insulation and then a lead sheath is placed. Outside diameter of this cable is nearly three inches, and the weight nine pounds per linear foot.

The 11,000-volt, three-phase current from Niagara Falls is distributed from the terminal house to seven sub-stations in Buffalo through about 30 miles of rubber-insulated and 18 miles of paper-insulated, three-conductor, lead-covered cables, all in tile conduits. In each cable the three No. 000 stranded conductors are separately insulated and then twisted into a rope with jute yarn laid in to give an even round surface for the lead sheath to rest on. A part of the rubber-insulated cables have each conductor covered with 9/32-inch of 30 per cent pure rubber compound, and the remaining rubber cables have 8/32-inch covering on each conductor of 40 per cent pure rubber compound. The paper-insulated cable has 13/64-inch of paper around each conductor, and also another 13/64-inch of paper covering about the group of three conductors and next to the lead sheath. In outside diameter the rubber-insulated cable is 2⅜ inches, and of the paper-insulated cable 2⅝ inches, the radial thickness of the lead sheath being ⅛-inch in each case. It is reported that the cables insulated with 9/32-inch of the mixture, said to be 30 per cent pure rubber, have proved to be more reliable than the cables insulated with 8/32-inch of a mixture said to be 40 per cent pure rubber. Vol. xviii., A. I. E. E., 136, 836.

The six miles of underground cables that carry three-phase, 25,000-volt current in St. Paul are of the three-conductor type, lead covered, and laid in a tile conduit. One of the two three-mile cables is insulated with rubber and the other with paper. In the former cable each conductor is separately insulated with a compound containing about 35 per cent of pure rubber and having a radial thickness of 7/32-inch. The three conductors after being insulated are laid up with jute to give a round surface, tape being used to hold them together, and then a rubber cover 5/32-inch thick is placed about the group, after which comes the lead sheath over all. In the three miles of paper-insulated cable each conductor is separately covered with paper to a thickness of $\frac{9}{32}$-inch, then the three conductors are laid together with jute and taped, and next a layer of paper $\frac{4}{32}$-inch thick is put on over the group. Outside of all comes the lead sheath, which has an outer coating of tin. The paper insulation in these cables was saturated with a secret insulating compound. The lead sheath on both the rubber and paper insulated cables is $\frac{1}{8}$-inch thick and the sheath of the former contains 3 per cent of tin. Each of the three conductors in each cable consists of 7 copper strands and has an area of 66,000 circular mils. Outside of the lead sheath each of these cables has a diameter of about 2$\frac{1}{4}$ inches. By the manufacturer's contract these cables were tested up to 40,000 volts before shipment, and might be tested up to 30,000 volts in their conduits during any time within five years from their purchase. In first cost the cable with rubber insulation was said to be about 50 per cent more expensive than the cable in which paper was used. Vol. xvii., A. I. E. E., 650.

Underground cables in which the separate conductors are covered with cotton braid treated with an insulating compound, and then the group of conductors going to make up the cable enclosed in a lead sheath, are extensively used in Austria and Germany. For cables that operate at 10,000 to 12,000 volts the radial thickness of cotton insulation on each conductor is said to be within $\frac{3}{16}$-inch, and these cables are tested up to 25,000 volts by placing all of the cable except its ends in water, and then connecting one end of the 25,000-volt circuit to the water and the other end to the conductors of the cable.

A test on the paper-insulated cable at St. Paul showed its charging current to be 1.1 amperes at 25,000 volts for each mile of its length. For the cable with rubber insulation the charging current per mile of length was found to be about twice as great as the like current for the paper-insulated cable. Each of the two overhead transmission lines connected with these cables consisted of three solid copper wires with an area of 66,000 circular mils each, and all three so mounted on the poles as to form the corners of an equilateral triangle twenty-four inches apart. The charging current of one of these three-wire, overhead circuits was found to be about 0.103 ampere per mile, at 25,000 volts, or a little less than one-tenth of the like current for the paper cable. These tests were made with three-phase current of sixty cycles per second.

Where overhead transmission lines join underground or submarine cables, either with or without the intervention of transformers, lightning arresters should be provided to intercept discharges of this sort that come over the overhead wires. Lightning arresters were provided in the terminal house at Buffalo, where the 22,000-volt overhead lines feed the 11,000-volt cables through transformers, also at the terminal house in St. Paul, where the

25,000-volt overhead lines are electrically connected to the underground cables. If an underground or submarine cable connects two portions of an overhead line, as in the Portsmouth and Dover transmission above mentioned, lightning arresters should be provided at each end of the cable, as was done in that case. One advantage of a high rather than a low voltage on underground cables, where power is to be transmitted at any given rate, lies in the fact that the amperes flowing at a fault in the cable determine the destructive effect there, rather than the voltage of the transmission. It is reported that a fault or short-circuit in one of the 11,000-volt cables at Buffalo usually melts off but little lead at the sheath and does not have enough explosive force to injure the cable or its duct.

Ozone seems to destroy the insulating properties of rubber very rapidly, and as it is well known that the silent electric discharge from conductors at high voltages develops ozone, care should be taken to protect rubber insulation from its action. This is especially true at the ends of cables where connections are made with switches or other apparatus, and the rubber insulation is exposed. To protect the rubber at such points it is the practice to solder a brass cable head or terminal bell to the lead sheath near its end, this head having a diameter perhaps twice as great as that of the sheath, and then to fill the space about the cable conductors in this head with an insulating compound. Heads of this sort were used on the 11,000-volt cables at Buffalo as well as on the 13,500-volt cable in the Portsmouth and Dover transmission.

As insulating materials, whether rubber, cotton, or paper, may be impaired or destroyed by heat, it is necessary that the temperature of underground cables under full load be kept within safe limits. Rubber insulation can probably be raised to 125° or 150° Fahrenheit without injury, and paper and cotton may go a little higher. For a given size and make of leaded cable the rise of temperature in its conductors above that of the surrounding air, for a given loss in watts per foot of the cable, may be determined by computation or experiment. The next step is to find out how much the temperature of the air in the conduits where the cable is to be used will rise above the temperature of the earth in which the conduits are laid, with the given watt loss per foot of cable. On this point there are but little experimental data. Obviously, the material of which ducts are made, the number of ducts grouped together with cables operating at the same time, and the extent to which ducts are ventilated must have an important bearing on this question. At Niagara Falls some tests were made to show the rise of air temperature in a section of thirty-six-duct conduit lying between two manholes about 140 feet apart. For the purpose of this test twenty-four of the thirty-six ducts in the conduit had one No. 6 drawing-in wire passed through each of them. These twenty-four wires were connected into three groups of eight wires each, so that one group was all in ducts next to the surrounding earth, another group was one-half in ducts next to the earth and the other half in ducts separated from the earth by at least one duct, while the third group of wires was entirely in ducts separated from the earth by at least one duct. It was found that when enough current was sent through these wires to represent a loss of 5.5 watts per foot of ducts in which they were located, the rise of temperature in the air of the ducts next to the earth was about 108° Fahrenheit above that of the earth. For the ducts separated from the earth by at least one other duct the rise in temperature of contained air was 144° Fahrenheit above the earth. If the earth about the ducts reached 70° in hot weather, the temperature of air in the inner ducts, with a loss of 5.5 watts per duct foot, would thus be 214°. This temperature is too high for either rubber, cotton, or paper insulation, to say nothing of the amount by which the temperature of the conductors and insulation of a cable in operation must exceed that of the surrounding air in its duct. The cables actually installed in the ducts just considered were designed for a loss of 2.34 watts per foot. As the No. 6 wire used in the test did not nearly fill each duct as a cable would do, it would be very interesting to know how much ventilation took place while the test was going on. Unfortunately, this point was not reported. Vol. xviii., A. I. E. E., 508.

CHAPTER XV.
MATERIALS FOR LINE CONDUCTORS.

Copper, aluminum, iron, and bronze are all used for conductors in long-distance electric transmissions, but copper is the standard metal for the purpose. An ideal conductor for transmission lines should combine the best electrical conductivity, great tensile strength, a high melting point, low coefficient of expansion, hardness, and great resistance to oxidation. No one of the metals named possesses all of these properties in the highest degree, and the problem is to select the material best suited to each case. Aluminum suffers very slightly by exposure to the weather, copper and bronze suffer a little more, while iron and steel wire are attacked seriously by rust.

Iron, copper, and bronze are all so hard that little or no trouble has occurred from wires of these metals cutting or wearing away at the points of attachment to insulators. Aluminum, on the other hand, is so soft that swaying of the wire may, in time, cause material wear at the supports, or it may be cut by tie wires. But lines of aluminum wire have not been in use long enough to determine how much trouble is to be expected from its lack of hardness.

A small coefficient of expansion is desirable in transmission wires, because the strain on the wire itself and on its supports varies rapidly with the amount of vertical deflection of each span, becoming greater as the deflection decreases. Taking the expansion of copper as unity, that of aluminum is 1.4; of bronze, 1.1; and of iron and steel, 0.7. From these figures it follows that iron and steel wires show the least variation in the amount of sag between supports, and aluminum wire shows the most.

Wrought iron melts at about 2,800°, steel at 2,700°, copper at 1,929°, bronze at about the same point as copper, and aluminum at 1,157° Fahrenheit. This low melting point of aluminum may prove a source of trouble by opening a line of that material where some foreign wire falls on it. This, according to a report, was illustrated at a sub-station on a 30,000-volt transmission line where a destructive arc was started at the switchboard. Not being able to extinguish the arc in any other way, a lineman threw an iron wire across the aluminum lines just outside of the sub-station, and these lines were immediately melted through by the iron wire, thus opening the circuit. The trouble may have warranted so desperate a remedy in this case; but, as a rule, it does not pay to cut a transmission line in order to get rid of a short circuit.

In the ordinary construction of transmission lines on land the tensile strength of wire is secondary in importance to its electrical conductivity, because supports can be spaced according to the strength of the conductor used. When large bodies of water must be crossed, tensile strength is a prime requirement. Thus a 142-mile line from Colgate to Oakland, in California, crosses the Straits of Carquinez in the form of steel cables, each seven-eighths of an inch in diameter and 4,427 feet long. Steel wire was selected for this long span, probably because it can be given a greater tensile strength than that of any other metal. Annealed iron wire has a tensile strength between 50,000 and 60,000 pounds per square inch. Steel wires vary all the way from 50,000 to more than 350,000 pounds per square inch in strength, but mild steel wire with a strength ranging from 80,000 to 100,000 pounds per square inch is readily obtained.

Soft copper shows a tensile strength between 32,000 and 36,000 pounds per square inch, and hard-drawn copper between 45,000 and 70,000 pounds, depending on the degree of hardness. Silicon-bronze wires vary in strength from less than 60,000 to more than 100,000 pounds per square inch, and phosphor-bronze has a tensile strength of about 100,000 pounds. Bronze wires, like those of most alloys, show a much wider range of strength than those of iron or copper.

In silicon-bronze wire the electrical conductivity decreases as the tensile strength increases. The tensile strength of aluminum wire is lower than that of any other used in transmission lines, being only about 30,000 pounds per square inch. Solid aluminum wires of large size have given trouble by breaking under strains well within their nominal strength, due probably to imperfections or twists. This trouble is now generally avoided by the use of aluminum cables.

In that most necessary property of a transmission line—conductivity—copper excels all other metals except silver. Taking the conductivity of soft copper wire at 100, the conductivity of hard-drawn copper is 98; that of silicon-bronze ranges from 46 to 98; that of aluminum is 60; of phosphor-bronze, 26; of annealed iron wire, 14; and of steel wire of 100,000 pounds tensile strength per square inch, 11. Copper wire, both soft and hard, as regularly made, does not vary more than one per cent from the standard, and aluminum and annealed iron wires also show high uniformity as to resistance. Silicon-bronze and steel wires, on the other hand, fluctuate much in electrical conductivity. For any particular transmission line the resistance is usually determined by considerations apart from the metal to be used as a conductor, so that a line of given resistance or conductivity must be constructed of that material which best conforms to the requirements as to size of wire, weight, strength, and cost.

Allowing the weight of any definite mass of copper to represent unity, the weight of an equal mass of wrought iron is 0.87; of steel, 0.89; of aluminum,

0.30; while that of bronze is very nearly equal to that of the copper. The smallest line wire that can be used for a given length and resistance is one of pure, soft copper. Next in cross-sectional area come hard-drawn copper and some silicon-bronze, either of which need be only two per cent larger than the soft copper for an equal resistance. Some other silicon-bronze wire of greater tensile strength per square inch would require a sectional area of 2.17 times that of the soft copper.

Aluminum wire with 60 per cent of the conductivity of copper requires 1.66 of its section for wires of equal resistance. As phosphor-bronze has only 26 per cent of the conductivity of copper, the section of the bronze must be 3.84 times that of the copper wire if their lengths and resistance are to be equal. An annealed iron wire is equal in resistance to a copper wire of the same length when the iron has 7.14 times the section of the copper. Steel, with 11 per cent of the conductivity of copper, must have 9.09 times the copper section in order that wires of the same length may have equal resistances.

It is not desirable to use a copper wire smaller than No. 4 B. & S. gauge for transmission lines, because of the lack of tensile strength in smaller sizes. When the conductivity of a copper wire smaller than No. 4 is ample, an iron wire will give the required conductivity, with a strength far greater than that of the copper. For a line of given length and conductivity of any other metal the weight compared with that of a copper line is represented by the product of the figures for relative section of the two lines and of the weight of unit mass of the metal in question compared with that of copper.

Thus, for the same conductivity the weight of a certain length of iron wire is $0.87 \times 7.14 = 6.21$ times the weight of a copper wire. For the steel wire above named the weight is $0.89 \times 9.09 = 8.09$ times that of a copper line of equal conductivity. Phosphor-bronze in a line of given length and resistance has 3.84 times the weight of soft copper. Silicon-bronze for a transmission line must weigh from 1.02 to 2.17 times as much as soft copper for a given length and conductivity. Aluminum for a line of fixed length and conductivity will weigh $1.66 \times 0.3 = 0.5$ times as much as copper. For a line of fixed length and resistance, hard-drawn copper will weigh about two per cent more than soft copper.

Taking the tensile strength of soft copper at 34,000 pounds per square inch, hard-drawn copper at 45,000 to 70,000, silicon-bronze at 60,000 to 100,000, phosphor-bronze at 100,000, iron at 55,000, steel at 100,000, and aluminum at 30,000 pounds, the relative strengths of wires with equal sectional areas compared with the soft copper are, for hard-drawn copper, 1.32 to 2.06; silicon-bronze, 1.76 to 2.94; phosphor-bronze, 2.94; iron, 1.62; steel, 2.94; and for aluminum, 0.88.

Comparing wires on the basis of equal resistances for equal lengths, with soft copper again the standard, the tensile strength of each as to it is as follows: A hard-drawn copper line has $1.02 \times 1.32 = 1.34$ to $1.02 \times 2.06 = 2.10$ times the strength of a line of soft copper. With silicon-bronze the strength of line wire would range between $1.02 \times 1.76 = 1.79$ and $2.17 \times 2.94 = 6.38$ times that of copper. Iron would give the line a strength as to soft copper represented by $7.14 \times 1.62 = 11.56$. Steel of 100,000 pounds tensile strength per square inch will give a line $9.09 \times 2.94 = 26.70$ times as strong as it would be if composed of soft copper. With aluminum the strength of the line would be $1.66 \times 0.88 = 1.46$ times that of copper. For phosphor-bronze the figures are $3.84 \times 2.94 = 11.29$.

From the foregoing it may be shown how many times the price of soft copper per pound may be paid for each of the other metals to form a line of given length and resistance at a cost equal to that of a soft copper line. These prices per pound for the several metals relative to that of soft copper are as follows: Taking the price of soft copper as one, the price for hard-drawn copper must be $1 \div 1.02 = 0.98$. For silicon-bronze the price may be as high as $1 \div 1.02 = 0.98$, or as low as $1 \div 2.17 = 0.46$ of the price of soft copper wire. Phosphor-bronze may have a price represented by only $1 \div 3.84 = 0.26$ that of copper. The price of iron wire should be $1 \div 6.21 = 0.16$ of that of copper, and for steel wire of the quality stated the price can only be $1 \div 8.01 = 0.12$. Aluminum wire alone may have a higher price per pound than soft copper for the same resistance and cost of line, the figure for the relative cost of this metal being $1 \div 0.5 = 2$.

From the foregoing it appears that for a line of given cost, length, and resistance, soft copper has the least cross-section and tensile strength; steel, the greatest cross-section, weight, tensile strength, and lowest permissible price per pound; and aluminum, the least weight and highest price per pound.

Relative Properties of Wires Having Equal Lengths and Resistances.

Metal in Wire.	Relative Cross Sections.	Relative Weights.	Relative Tensile Strengths.	Relative Prices per Pound for Same Total Cost.
Soft Copper	1.00	1.00	1.	1.00
Hard Copper	1.02	1.04	1.34	.98
Very Hard Copper	1.02	1.02	2.10	.98
No. 1 Silicon-Bronze	1.02	1.02	1.79	.98
No. 2 Silicon-Bronze	2.17	2.17	6.38	.46
Aluminum	1.66	.50	1.46	2.00
Phosphor-Bronze	3.84	3.84	11.29	.26
Annealed Iron	7.14	6.21	11.56	.16
Mild Steel	9.09	8.09	26.70	.12

The relative cross sections and weights of both iron and steel wires are so great as to prevent their general use because of the labor and cost of their erection.

So far as the first cost of the wire alone is concerned, iron may be approximately equal to copper in some metal markets. The only practical place for an iron wire, however, is one where copper would be too small or not strong enough. Steel wire finds a place, in spite of its high resistance, in those exceptional cases where a single span of several thousand feet must be made, requiring high tensile strength. In such cases it is usually better to give the steel span a greater resistance than an equal length of the main portion of the line, so as to avoid excessive size and weight of the span. Even when this is done the resistance of the steel span would be very small compared with that of a long transmission line.

Phosphor-bronze finds little use as conductors in transmission systems because of its relatively high electrical resistance. If great tensile strength is wanted, iron or steel will supply it at a fraction of the cost of phosphor-bronze. As a conductor simply, phosphor-bronze is worth only 0.26 as much per pound as soft copper, while its actual market price is greater than that of copper.

Silicon-bronze of relatively high resistance, requiring 2.17 times the section and weight of copper for equal conductivity, is entitled to little or no consideration as a transmission line material. This alloy, in order to give equal conductivity at equal cost with copper, must sell at only 0.46 of the price of copper per pound. But the price of silicon-bronze is equal to, or greater than, the price of copper, so that the cost of the high-resistance silicon-bronze for a line of given resistance will be more than twice that of copper. For this more than double cost the bronze gives 6.38 times the tensile strength of a soft copper line of equal conductivity.

Taking the market price of steel at one-fifth that of copper, which is amply high for the steel, as a rule, a steel wire of equal conductivity with the copper will cost only 1.6 times as much and will have 26.7 times the tensile strength of the copper, or four times the tensile strength of a wire of equal conductivity made from the high-resistance silicon-bronze. From this it is clear that steel offers a cheaper combination of conductivity and strength than does silicon-bronze of high resistance. That grade of silicon-bronze having the lowest resistance may cost 0.98 as much per pound as soft copper, and will have 1.79 times the strength of the copper for equal conductivity. This bronze actually costs more per pound than copper, so that it cannot give equal conductivity at equal cost.

Very hard-drawn copper has a conductivity equal to that of the best silicon-bronze, and the tensile strength of this copper is seventeen per cent greater than that of the bronze. Silicon-bronze costs more per pound than hard copper, but even with equal prices the hard copper gives equal conductivity and higher strength at the same cost. Furthermore, the conductivity of silicon-bronze is much more liable to serious variations than that of hard copper. Between hard-drawn copper and steel there is very little apparent place for any grade of bronze in electric transmission lines.

The hardest copper wire is very stiff, and is more liable to crack when twisted or bent than is wire of only medium hardness. Such medium-hard copper has a tensile strength of thirty-four per cent greater than soft copper of equal conductivity, and is much used on long transmission lines. Aluminum is the only metal which, for given conductivity in a transmission line, combines a smaller weight, a greater tensile strength, and a higher permissible price than soft copper for the same total cost. For equal conductivity an aluminum wire has a greater tensile strength than one of medium-hard copper, and costs less than copper of any grade when the price per pound of the aluminum is less than twice that of copper, which is usually the case.

These properties make aluminum by far the most important competitor of copper in electric transmission and have led to its use in a number of cases, notably for the two longest lines in the world, namely, between Colgate and Oakland and between Electra and San Francisco, in California.

It has not been found practicable to solder joints in aluminum wires because of the resulting electrolytic action when aluminum is in contact with other metals. Joints of aluminum wires are usually made by slipping the ends past each other in an oval aluminum sleeve and then giving the sleeve and wires two or three complete twists, or by a process of cold welding with a sleeve joint.

Long transmission lines are in nearly all cases run with bare wire supported by poles. Where very high voltages are employed no insulation that can be put on the wire will make it safe to handle, and the cost of such insulation would add materially to that of the entire line. It is, therefore, the practice to run transmission lines above all other wires and to rely entirely on the supports for insulation.

The considerations thus far noted apply alike to wires carrying continuous and alternating currents, but there are some other factors that apply solely to alternating lines. Owing to the inductive effects of alternating currents in long, parallel wires, such wires should be transposed between their supports at frequent intervals. The induction between wires increases with the frequency of the current carried, and decreases with the distance between the wires. According to these conditions, wires should be transposed as often as every eighth of a mile in some cases, and at intervals of one mile or more in others.

An alternating current when passing along a line tends to concentrate itself in the outer layers of the wire, leaving the centre idle. This unequal current distribution increases with the frequency of the current and with the area of the cross section of the wire. The practical effect of this unequal distribution is to make the resistance of a wire a little higher for alternating than for continuous currents. In existing transmission lines the increase of resistance due to this cause seldom amounts to one per cent.

When an alternating current passes through a circuit, the action termed self-induction sets up an electromotive force in the circuit that opposes the flow of current, as does the resistance of the wire, and this is called the inductance of the circuit. The ratio of this inductance to the resistance of a circuit increases with the number of periods per second of the alternating current used and with the sectional area of the wires composing the circuit. For a circuit of No. 6 B. & S. gauge wire the inductance amounts to only five per cent of the line resistance, but for a circuit of No. 000 wire the inductance consumes as much of the applied voltage as does the resistance, with 60-cycle current.

Both the unequal distribution of alternating current over the cross-section of a conductor and the inductance of circuits make it desirable to keep the diameters of transmission wires as small as other considerations permit. As soft copper has greater conductivity per unit of area than any of the other available metals, it clearly has an advantage over all of them as to inductance and increase of resistance with alternating current.

At very high voltages there is an important leakage of energy between the conductors of a circuit, and this loss varies inversely with the distance between these conductors. Thus it happens that inductance makes it desirable to bring the parallel wires of a circuit close together, while the leakage of energy from wire to wire makes it desirable to carry them far apart.

To provide greater security from interruption, the conductors for important transmissions are in some cases carried on two independent pole lines. Even where all the conductors are on a single line of poles it is frequent practice to divide them up into a number of comparatively small wires, and this decreases inductance.

Data of a number of transmission lines presented in the appended table illustrate the practice in some of the more recent and important cases as to the materials, size, number, and arrangement of the wires. The plants of which particulars are given include the greatest power capacities, the longest distances, and the highest voltages now involved in electrical transmissions. Each of the lines named is worked with alternating current of two- or three-phase. Each three-phase line must have at least three wires, and each two-phase line usually has four wires.

On ten of the lines the number of wires is greater than three or four, thus reducing the necessary size of each wire for a given conductivity of the line. The Butte, Oakland, and Hamilton lines are run on two sets of poles for greater security, and a second pole line has been added to the Niagara and Buffalo system to carry additional wires.

The largest wire used in any of these lines is the aluminum cable of 500,000 circular mils between Niagara Falls and Buffalo. This cable has 1.66 times the area in cross section of a copper cable of equal conductivity.

SIZES AND MATERIALS OF WIRES ON SOME AMERICAN TRANSMISSION LINES.

Location of Transmission.	Line Voltage.	Number Wires.	Size of Each Wire B. & S. Gauge.	Metal in Wire.	Length of Transmission. Miles.
Cañon Ferry to Butte	50,000	6	0	Copper	65
Colgate to Oakland	40,000	3	00	Copper	142
		3	000	Aluminum	142
Electra to San Francisco	40,000	3	471,034 C. M.	,,	147
Santa Ana R. to Los Angeles	33,000	6	1	Copper	83
Apple River to St. Paul	25,000	6	2	,,	25
Welland Canal to Hamilton	22,500	3	1	,,	35
		3	00	,,	37
Cañon City to Cripple Creek	20,000	3	3	,,	23 ½
Madrid to Bland	20,000	6	4	,,	32
White River to Dales	22,000	3	6	,,	27
Ogden to Salt Lake City	16,000	6	1	,,	36 ½
San Gabriel Cañon to Los Angeles	16,000	6	5	,,	23
To Victor, Col	12,600	3	4	,,	8
Niagara Falls to Buffalo	22,000	6	350,000 C. M.	,,	23
Nia,,ara Fa,,s to Bu,,	22,000	3	500,000 C. M.	Aluminum	20
Yadkin River to Salem	12,000	3	1	Copper	14.5
Farmington Riv'r to Hartford	10,000	3	336,420 C. M.	Aluminum	11
Wilbraham to Ludlow Mills	11,500	6	135,247 C. M.	,,	4.5
Niagara Falls to Toronto	60,000	6	190,000 C. M.	Copper	75

Aluminum lines are now employed for the three longest electrical transmissions in North America. In the longest single line, that from Electra power-house to San Francisco, a distance of 147 miles, aluminum is the conductor used. The 142-mile transmission between Colgate and Oakland is carried out with three aluminum and three copper line wires. For the third transmission in point of length, that from Shawinigan Falls to Montreal, a distance of 85 miles, three aluminum conductors are employed.

The three transmissions just named have unusually large capacities as well as superlative lengths, the generators in the Electra plant being rated at 10,000, in the Colgate plant at 11,250, and in the Shawinigan plant at 7,500 kilowatts. Weight and cost of such lines are very large. For the three No. 0000 aluminum conductors, 142 miles each in length, between Colgate and Oakland, the total weight must be about 440,067 pounds, costing $132,020 at 30 cents per pound. Between Electra and Mission San José, where the line branches, is 100 miles of the 147-mile transmission from Electra to San Francisco. On the Electra and Mission San José section the aluminum conductors comprise three stranded cables of 471,034 circular mils each in sectional area and with a total weight of about 721,200 pounds. This section alone of the line in question would have cost $216,360 at 30 cents per pound. The 85-mile aluminum line from Shawinigan Falls to Montreal is made up of three-stranded conductors each with a sectional area of 183,708 circular mils. All three conductors have a combined weight of about 225,300 pounds, and at 30 cents per pound would have cost $67,590.

Aluminum lines are not confined to new transmissions, but are also found in additions to those where copper conductors were at first used. Thus, the third transmission circuit between the power-house at Niagara Falls and the terminal house in Buffalo, a distance of 20 miles by the new pole line, was formed of three aluminum cables each with an area of 500,000 circular mils, though the six conductors of the two previous circuits were each 350,000 circular mils copper.

From these examples it may be seen that copper has lost its former place as the only conductor to be seriously considered for transmission circuits. Aluminum has not only disputed this claim for copper, but has actually gained the most conspicuous place in long transmission lines. This victory of aluminum has been won in hard competition. The decisive factor has been that of cost for a circuit of given length and resistance.

From the standpoint of cross-sectional area aluminum is inferior to copper as an electrical conductor. Comparing wires of equal sizes and lengths, the aluminum have only sixty per cent of the conductivity of the copper, so that an aluminum wire must have 1.66 times the sectional area of a copper wire of the same length in order to offer an equal electrical resistance. As round wires vary in sectional areas with the squares of their diameters, an aluminum wire must have a diameter 1.28 times that of a copper wire of equal length in order to offer the same conductivity.

The inferiority of aluminum as an electrical conductor in terms of sectional area is more than offset by its superiority over copper in terms of weight. One pound of aluminum drawn into a wire of any length will have a sectional area 3.33 times as great as one pound of copper in a wire of equal length. This follows from the fact that the weight of copper is 555 pounds while that of aluminum is only 167 pounds per cubic foot, so that for equal weights the bulk of the latter is 3.33 times that of the former metal. As the aluminum wire has equal length with and 3.33 times the sectional area of the copper wire of the same weight, the electrical conductivity of the former is $3.33 \div 1.66 = 2$ times that of the latter. Hence, for equal resistances, the weight of an aluminum is only one-half as great as that of a copper wire of the same length. From this fact it is evident that when the price per pound of aluminum is anything less than twice the price of copper, the former is the cheaper metal for a transmission line of any required length and electrical resistance.

The tensile strength of both soft copper and of aluminum wire is about 33,000 pounds per square inch of section. For wires of equal length and resistance the aluminum is therefore sixty-six per cent stronger because its area is sixty-six per cent greater than that of a soft copper wire. Medium hard-drawn copper wire such as is most commonly used for transmission lines has a tensile strength of about 45,000 pounds per square inch, but even compared with this grade of copper the aluminum wire of equal length and resistance has the advantage in tensile strength. While the aluminum line is thus stronger than an equivalent one of copper, the weight of the former is only one-half that of the latter, so that the distance between poles may be increased, or the sizes of poles, cross-arms, and pins decreased with aluminum wires. In one respect the strain on poles that carry aluminum may be greater than that on poles with equivalent copper lines, namely, in that of wind pressure. A wind that blows in a direction other than parallel with a transmission line tends to break the poles at the ground and prostrate the line in a direction at right angles to its course. The total wind pressure in any case is obviously proportional to the extent of the surface on which it acts, and this surface is measured by one-half of the external area of all the poles and wires in a given length of line. As the aluminum wire must have a diameter twenty-eight per cent greater than that of copper wire of equal length, one-half of the total wire surface will also be twenty-eight per cent greater for the former metal. This carries with it an increase of twenty-eight per cent in that portion of the wind pressure due to wire surface. In good practice the number of transmission wires per pole line is often only three, and seldom more than six, so that the surface areas of these wires may be no greater than that of the poles. It follows that the increase of twenty-eight per cent in the surface of wires may correspond to a much smaller percentage of increase for the entire area exposed to wind pressure. Such small difference as exists between the total wind pressures on aluminum and copper lines of equal conductivity is of slight importance in view of the general practice by which some straight as well as the curved portions of transmission lines are now secured by guys or struts at right angles to the direction of the wires.

Vibration of transmission lines and the consequent tendency of cross-arms, pins, insulators, and of the wires to work loose is less with aluminum than with copper conductors as ordinarily strung, because of the greater sag between poles given the former and also probably because of their smaller weight. An illustration of this sort may be seen on the old and new transmission lines between Niagara Falls and Buffalo. The two old copper circuits consist of six cables of 350,000 circular mils section each on one line of poles, and are strung with only a moderate sag. In a strong wind these copper conductors swing and vibrate in such a way that their poles, pins, and cross-arms are thrown into a vibration that tend to work all attachments loose. The new circuit consists of three 500,000 circular mil aluminum conductors on a separate pole line strung with a large sag between poles, and these conductors take positions in planes at large angles with the vertical in a strong wind, but cause little or no vibration of their supports. One reason for the greater sag of the aluminum over that of the copper conductors in this case is the fact that the poles carrying the former are 140 feet apart while the distance between the poles for the latter is only seventy feet, on straight sections of the line.

The necessity for greater sag in aluminum than in copper conductors, even where the span lengths are equal, arises from the greater coefficient of expansion possessed by the former metal. Between 32° and 212° Fahrenheit aluminum expands about 0.0022, and copper 0.0016 of its length, so that the change in length is 40 per cent greater in the former than in the latter metal. The conductors in any case must have enough sag between poles to provide for contraction in the coldest weather, and it follows that the necessary sag of aluminum wires will be greater than that of copper at ordinary temperature.

In pure air aluminum is even more free from oxidation than copper, but where exposed to the fumes of chemical works, to chlorine compounds, or to fatty acids the metal is rapidly attacked. For this reason aluminum conductors should have a water-proof covering where exposed to any of these chemicals. The aluminum line between Niagara Falls and Buffalo is bare for most of its length, but in the vicinity of the large chemical works at the former place the wires are covered with a braid treated with asphaltum. Aluminum alloyed with sodium, its most common impurity, is quickly corroded in moist air, and should be carefully avoided. All of the properties of aluminum here mentioned relate to the pure metal unless otherwise stated, and its alloys should not, as a rule, be considered for transmission lines. As aluminum is electropositive to most other metals the soldering of its joints is quite sure to result in electrolytic corrosion, unless the joints are thoroughly protected from moisture, a result that is hard to attain with bare wires. The regular practice is to avoid the use of solder and rely on mechanical joints. A good joint may be made by slipping the roughened ends of wires to be connected through an aluminum tube of oval section, so that one wire sticks out at each end, then twisting the tube and wires and giving each of the latter a turn about the other. Aluminum may be welded electrically and also by hammering at a certain temperature, but these processes are not convenient for line construction. Especial care is necessary to avoid scarring or cutting into aluminum wires, as may be done when they are tied to their insulators. Aluminum tie wires should be used exclusively. To avoid the greater danger of damage to solid wires and also to obtain greater strength and flexibility, aluminum conductors are most frequently used in the form of cables. The sizes of wires that go to make up these cables commonly range from No. 6 to 9 B. & S. gauge for widely different cable sections. Thus the 183,708 circular mil aluminum cable between Shawinigan Falls and Montreal is made up of seven No. 6 wires, and the 471,034 circular mil cable between Electra and Mission San José contains thirty-seven No. 9 wires. From the Farmington River to Hartford each 336,420 circular mils cable has exceptionally large strands of approximately No. 3 wire. It appears from the description of a 43-mile line in California (vol. xvii., A. I. E. E., p. 345) that a solid aluminum wire of 294 mils diameter, or No. 1 B. & S. gauge, can be used without trouble from breaks. This wire was tested and its properties reported as follows:

- Diameter, 293.9 mils.
- Pounds per mile, 419.4.
- Resistance per mil foot, 17.6 ohms at 25° C.
- Resistance per mile at 25° C., 1.00773 ohms.
- Conductivity as to copper of same size, 59.9 per cent.
- Number of twists in six inches for fracture, 17.9.
- Tensile strength per square inch, 32,898 pounds.

This wire also stood the test of wrapping six times about its own diameter and then unwrapping and wrapping again. It was found in tests for tensile strength that the wire in question took a permanent set at very small loads, but that at points between 14,000 and 17,000 pounds per square inch the permanent set began to increase very rapidly. From this it appears that aluminum wires and cables should be given enough sag between poles so that in the coldest weather the strains on them shall not exceed about 15,000 pounds per square inch. This rather low safe working load is a disadvantage that aluminum shares with copper. From the figures just given it is evident that the strains on aluminum conductors during their erection should not exceed one-half of the ultimate strength in any case, lest their sectional areas be reduced.

ALUMINUM CABLES IN TRANSMISSION SYSTEMS.

Locations.	Number of Cables.	Miles of Each.	Circular Mils of Each.	Strands per Cable.	Size of Strands. B. & S. G. Approximate.
Niagara Falls to Buffalo	3	20	500,000	..	..
Shawinigan Falls to Montreal	3	85	183,708	7	6
Electra to Mission San José	3	100	471,034	37	9
Colgate to Oakland	3	144	211,000	7	5-6
Farmington River to Hartford	3	11	336,420	7	3
Lewiston, Me.	3	3.5	144,688	7	8
Ludlow, Mass.	6	4.5	135,247	7	7

This table of transmission systems using aluminum conductors is far from exhaustive. Aluminum is also being used to distribute energy to the sub-stations of long electric railways, as on the Aurora and Chicago which connects cities about forty miles apart. The lower cost of aluminum conductors is also leading to their adoption instead of copper in city distribution of light and power. Thus at Manchester, N. H., the local electric lines include about four miles each of 500,000 and 750,000 circular mil aluminum cable with weather-proof insulation. The larger of these cables contains thirty-seven strands of about No. 7 wire.

As may be seen from the foregoing facts, the choice of copper or aluminum for a transmission line should turn mainly on the cost of conductors of the required length and resistance in each metal. So nearly balanced are the mechanical and electrical properties of the two metals that not more than a very small premium should be paid for the privilege of using copper. As already pointed out, the costs of aluminum and copper conductors of given length and resistance are equal when the price per pound of aluminum wire is twice that of copper. During most of the time for several years the price of aluminum has been well below double the copper figures, and the advantage has been with aluminum conductors. With the two metals at the same price per pound aluminum would cost only one-half as much as equivalent copper conductors. When the price of aluminum is fifty per cent greater per pound than that of copper, the use of the former metal effects a saving of twenty-five per cent. For the new Niagara and Buffalo line, completed early in 1901, aluminum was selected because it effected a saving of about twelve per cent over the cost of copper. All of the aluminum lines here mentioned, except the short one near Hartford, were completed during or since 1900. Most of the facts here stated as to the line between Niagara Falls and Buffalo are drawn from vol. xviii., A. I. E. E., at pages 520 and 521.

The greater diameter of aluminum over equivalent copper conductors has advantages in transmission with alternating current at very high voltages. At high voltages, say of 40,000 or more, the constant silent loss of energy from one conductor to another of the same circuit through the air tends to become large and even prohibitive in amount. This loss is greater, other factors being constant, the smaller the diameter of the conductors in the line. It follows that this loss is more serious the smaller the power to be transmitted, because the smaller the diameter of the line wires. The silent passage of energy from wire to wire increases directly with the length of line and thus operates as a limit to long transmissions.

CHAPTER XVI.
VOLTAGE AND LOSSES ON TRANSMISSION LINES.

The voltage on a transmission line may be anything up to at least 60,000, and the weight of conductors varies inversely with the square of the figures selected, the power, length and loss being constant. Whatever the total line pressure, the weight of conductors varies inversely with the percentage of loss therein.

The case of maximum loss and minimum weight of conductors is that in which all of the transmitted energy is expended in heating the line wires. Such a case would never occur in practice, because the object of power transmission is to perform some useful work.

Minimum loss is theoretically zero, and the corresponding weight of conductors is infinite, but these conditions obviously cannot be attained in practice. Between these extremes of minimum and of infinite weights of conductors comes every practical transmission with a line loss greater than zero and less than 100 per cent.

To determine the weight and allowable cost of conductors, the cost of the energy that will be annually lost in them enters as one of the factors. At this point the distinction between the percentage of power lost at maximum load and the percentage of total energy lost should come into view.

Line loss ordinarily refers to the percentage of total power consumed in the conductors at maximum load. This percentage would correspond with that of total energy lost if the line current and voltage were constant during all periods of operation, but this is far from the case.

A system of transmission may operate with either constant volts or constant amperes on the line conductors, but in a practical case constancy of both these factors is seldom or never to be had. This is because the product of the line volts and amperes represents accurately in a continuous-current system, and approximately in an alternating-current system, the amount of power transmitted. In an actual transmission system, the load—that is, the demand for power—is subject to more or less variation at different times of the day, and the line volts or amperes, or both, must vary with it.

If the transmission system is devoted to the operation of one or more factories the required power may not vary more than twenty-five per cent during the hours of daily use; but if a system of general electrical supply is to be operated, the maximum load will usually be somewhere between twice and four times as great as the average load for each twenty-four hours. Such fluctuating loads imply corresponding changes in the volts or amperes of the transmission line.

A number of rather long transmissions is carried out in Europe with continuous, constant current, and in such systems the line voltage varies directly with the load. As the loss of power in an electrical conductor depends entirely on its ohms of resistance, which are constant at any given temperature, and on the amperes of current passing through it, the line loss in a constant-current system does not change during the period of operation, no matter how great may be its changes of load. For this reason the percentage of power loss in the line at maximum load is usually smaller than the percentage of energy loss for an entire day.

If, for example, the constant-current transmission line is designed to convert into heat 5 per cent of the maximum amount of energy that will be delivered to it per second—that is, to lose 5 per cent of its power at maximum load—then, when the power which the line receives drops to one-half of its maximum, the percentage of loss will rise to 10, because $0.05 \div 0.5 = 0.1$. So again, when the power sent through the line falls to one-quarter of the full amount, the line loss will rise to $0.05 \div 0.25 = 0.2$, or 20 per cent.

From these facts it is clear that a fair all-day efficiency for a constant-current transmission line can be obtained only in conjunction with a high efficiency at maximum load, if widely varying loads are to be operated. It does not necessarily follow from these facts as to losses in constant-current lines that such losses should always be small at maximum loads, for if a large loss may be permitted at full load a still greater percentage of loss at partial loads may not imply bad engineering.

In a large percentage of electric water-power plants some water goes over the dam during those hours of the day when loads are light, the storage capacity above the dam not being sufficient to hold all of the surplus water during most seasons of the year. If, therefore, the line loss in a constant-current transmission, where all of the daily flow of water cannot be used, is not great enough to reduce the maximum load that would otherwise be carried, then the fact that the percentage of line loss at small loads is still larger is not very important.

Obviously, it makes little difference whether water goes over a dam or through wheels to make up for a loss in the line. In a case where all the water can be stored during small loads and used during heavy loads, it is clearly desirable to keep the loss in a constant-current line down to a rather low figure, say not more than five per cent, at maximum load.

Much the greater number of electrical transmissions are carried out with nearly constant line voltage, mostly alternating, and the line current in such cases varies directly with the power transmitted, except as to certain results of inductance on alternating lines. As line resistance is constant, save for slight variations due to temperature, the rate of energy loss on a constant-pressure line varies with the square of the number of amperes flowing, and the percentage of loss with any load varies directly as the number of amperes.

These relations between line losses and the amperes carried follow from the law that the power, or rate of work, is represented by the product of the number of volts by the number of amperes, and the law that the power actually lost in the line is represented by the product of the number of ohms of line resistance and the square of the number of amperes flowing in it. In each of these cases the power delivered to the line is, of course, measured in watts, each of which is $1/746$ of a horse-power.

Applying these laws, it appears that if the loss of a certain constant-pressure transmission line is 10 per cent of the power delivered to it at full load, then, when the power, and consequently the amperes, on the line is reduced one-half, the watts lost in the line as heat will be $(1/2)^2 = 1/4$ of the watts lost at full load, because the number of amperes flowing has been divided by 2.

But the amount of power delivered to the line at full load having been reduced by 50 per cent, while the power lost on the line dropped to one-fourth of 10 per cent, or to 2.5 per cent of the full line load, it follows that the power lost on the line at half-load is represented by $0.025 \div 0.5 = 0.05$, or 5 per cent of the power then delivered to it.

This rise in the efficiency of a constant-pressure transmission line as the power delivered to it decreases, together with the fact that maximum loads on such lines continue during hardly more than one to two hours daily, tends to raise the allowable percentage of line loss at maximum loads.

This is so because a loss of fifteen per cent at maximum load may easily drop to an average loss of somewhere between five and ten per cent for the entire amount of energy delivered to a line during each day under ordinary conditions in electrical supply. In the practical design of transmission lines, therefore, the sizes of conductors are influenced by the relation of the largest load to be operated to the greatest amount of power available for its operation, and by questions of regulation, as well as by considerations of all-day efficiency.

If the maximum load that must be carried by a transmission system during a single hour per day requires nearly as much power as can be delivered to the line conductors, either because of lack of water storage or of water itself, even if it is stored, it may be desirable to design these conductors for a small loss at maximum load, rather than to install a steam plant.

So again, as the fluctuation in voltage at the delivery end of a transmission line between no load and full load will amount to the entire drop of volts in the line at full load, if the pressure at the generating end is constant, the requirements of pressure regulation on distribution circuits limit the drop of pressure in the transmission conductors. For good lighting service with incandescent lamps at about 110 volts, the usual pressure, it is necessary that variations be held within one volt either way of the pressure of the lamps—that is, between 109 and 111 volts.

Every long-transmission system for general electrical supply necessarily includes one or more sub-stations where the distribution lines join the transmission circuits, and where the voltage for lighting service is regulated. As the limits of voltage variations on lighting circuits are so narrow, it is necessary to keep the changes of pressure on the transmission lines themselves within moderate limits, or such as can be compensated for at sub-stations.

This is particularly true in cases where energy transmitted over a single circuit is distributed for both incandescent lamps and large electric motors, because the starting and operation of such motors causes large fluctuations of amperes and terminal voltage on the transmission circuits. To hold such fluctuations within limits which a sub-station can readily compensate for, it is necessary that the loss in the transmission line be moderate, say often within ten per cent of the total voltage delivered to it at maximum load.

Capacity and cost of equipment at generating stations go up with the percentage of line loss, and thus serve to limit its economical amount. For every horse-power delivered to a transmission line at a water-power station there must be somewhat more than one horse-power of capacity in water-wheels, at least one horse-power in generators, and frequently a further capacity of one horse-power in step-up transformers. Every additional horse-power lost in the line at maximum load, if the generating plant is to be worked up to its full capacity, implies an addition of somewhat more than one horse-power capacity in water-wheels, one horse-power in generators, and one horse-power in transformers.

Since the cost of a generating station is thus increased as the maximum line loss is raised, a point may be reached where any further saving in the cost of the line is more than offset by the corresponding addition to the cost of the station and of its operation. Just where this point, as indicated by a percentage of line loss, is to be found depends on the factors of each case, important among which is the length of the transmission line.

Much effort has been made to fix some exact relation for maximum economy between the first cost of conductors for a transmission line and the amount of energy annually lost as heat therein. The best-known statement applying to this case is that of Lord Kelvin, made in a paper read before the British Association in 1881. According to the rule there laid down, the most economical size for the conductors of a transmission line is that for which the annual interest on first cost equals the cost of the energy annually wasted in them.

If transmission systems were designed for the sole purpose of wasting energy in their line conductors this rule would exactly apply, for it simply shows how the cost of energy wasted, plus the interest on the cost of the conductor in which it is wasted, may be brought to a minimum. As a matter of fact, transmission systems are primarily intended to deliver energy rather than to waste it; but of the proportions of the entire energy to be delivered and wasted (which is exactly what we want to know), the rule of Kelvin takes no account.

According to his rule, the cheaper the cost of power where it is developed, the less should be paid for conductors to bring it to market. The obvious truth is that the less the cost of power development at a particular point, the more may be invested in a line to bring it to market. If power cost nothing whatever at its source it would not be worth while to build any transmission line at all if this rule is correct.

A modification of Lord Kelvin's rule has been proposed by which it is said that the interest on the cost of the conductors and the annual value of the energy lost in them should be equal, value here meaning what the energy can be sold for. This rule would make an investment in line conductors too large.

The entire cost of production and transmission for the delivered energy should not be greater than the cost of a like amount of energy developed at the point where the delivery is made. In this entire cost of production and transmission, interest on the investment in line conductors is only one item.

It is perhaps impossible to state any exact rule for the most economical relation between the cost of conductors and the loss of energy therein that will apply to every transmission. A maximum limit to the weight of conductors may, however, be set for most cases. This limit should not allow the annual interest and depreciation charges on the investment in line conductors, plus all other costs of development and transmission, to raise the total cost of the transmitted energy above the cost of development for an equal amount of energy at the point where the transmitted energy is delivered.

While the maximum investment in transmission conductors may be properly limited in the way just stated, it by no means follows that this maximum limit should be reached in every case. In the varying requirements of actual cases, the problem may be to deliver a fixed amount of power at the least possible cost, or to deliver the largest possible amount of power at a cost per unit under that of development at the point of use. Frequently a transmission system has a possible capacity in excess of present requirements, and a line that would not be too heavy for future business might put an unreasonable burden of interest charges on present earnings.

The foregoing considerations apply to the design of conductors for a transmission line after the voltage at which it is to operate has been decided on. Quite a different set of facts should influence the selection of this voltage. A transmission that would be entirely impracticable with any percentage of line loss that might be selected, if carried out at some one voltage, might represent a paying business at some higher voltage and any one of several sizes of line conductors. The power that could be delivered by a line of practicable cost, operated at one voltage, might be too small for the purpose in hand, while the available power at a higher voltage might be ample.

If any given power is to be transmitted with a given percentage of maximum loss in line conductors, the weight of these conductors will increase as the square of their length, and decrease as the square of the full voltage of operation in every case.

Thus, if the length of this transmission is doubled, the weight of the conductors must be multiplied by four, the voltage remaining the same; but if the voltage is doubled and the line length remains unchanged, the weight of conductors must be divided by four. With the length of line and the voltage of transmission either lowered or raised together, the weight of the conductors remains fixed, for constant power and loss.

An illustration of this last rule may be drawn from the case of lines designed to transmit any given power a distance of ten miles at 10,000 volts, and a distance of fifty miles at 50,000 volts, in which the total weight of conductors would be the same for each line if the percentage of loss was constant.

This statement of the rule as to proportionate increase of voltage and distance presents the advantages of high voltages in their most favorable light. Though a uniform ratio between the voltage of operation and the length of line allows a constant weight of conductors to be employed for the transmission of a given power with unchanging efficiency of conductors, yet other considerations soon limit the advantage thus obtained.

Important among these considerations may be mentioned the mechanical strength of line conductors, difficulties of line insulation, losses between conductors through the air, limits of generator voltages, and the cost of transformers.

If the ten-mile transmission at 10,000 volts, above mentioned, requires a circuit of two No. 1/0 copper wires, the total weight of these wires will be represented by $(5{,}500 \times 10 \times 2 \times 320) \div 1{,}000 = 35{,}200$ pounds, allowing 5,500 feet of wire per mile of single conductor to provide something for sag between poles, and 320 pounds being the weight of bare No. 1/0 copper wire per 1,000 feet.

When the length of line is raised to 50 miles, the two-wire circuit will contain $5{,}500 \times 50 \times 2 = 550{,}000$ feet of single conductor, and since the voltage is raised to 50,000 at the same time, the total weight of conductors will be 35,200 pounds as before. The weight of single conductor per 1,000 feet is therefore only 64 pounds in the 50-mile line.

A No. 7 copper wire, B. & S. gauge, has a weight of 63 pounds per 1,000 feet, and is the nearest regular size to that required for the 50-mile line as just found. It would be poor policy to string a wire of this size for a transmission line, because it is so weak mechanically that breaks would probably be frequent in stormy weather. The element of unreliability introduced by the use of this small wire on a 50-mile line would cost far more in the end than a larger conductor.

As a rule, No. 4 B. & S. gauge wire is the smallest that should be used on a long transmission line in order to give fair mechanical strength, and this size has just twice the weight of a No. 7 wire of equal length. Here, then, is one of the practical limits to the advantages that may be gained by increasing the voltage with the length of line.

As line voltage goes up, the strain on line insulation increases rapidly, and the insulators for a circuit operated at 50,000 volts must be larger and of a much more expensive character than those for a 10,000-volt circuit. In this way a part of the saving in conductors effected by the use of very high voltages on long lines is offset by the increased cost of insulation.

Another disadvantage that attends the operation of transmission lines at very high voltages is the continuous loss of energy by the silent passage of current through the air between wires of a circuit. This loss increases at a rapid rate after a pressure between 40,000 and 50,000 volts is reached with ordinary distances between the wires of each circuit. To keep losses of this sort within moderate limits, and also to lessen the probability of arcs on a circuit at very high voltage, the distance of eighteen inches or two feet between conductors that carry current at 10,000 volts should be increased to six feet or more on circuits that operate at 50,000 volts.

Such an increase in the distance between conductors makes the cost of poles and cross-arms greater, either by requiring them to be larger than would otherwise be necessary or by limiting the number of wires to two or three per pole and thus increasing the number of pole lines. These added expenses form another part of the penalty that must be paid for the use of very high voltages and the attendant saving in the cost of conductors.

Apparatus grows more expensive as the voltage at which it is to operate increases, because of the cost of insulating materials and the room which they take up, thereby adding to the size and weight of the iron parts.

Generators for alternating current can be had that develop as much as 13,500 volts, but such generators cost more than others of equal power that operate at between 2,000 and 2,500 volts. These latter voltages are as high as it is usually thought desirable to operate distribution circuits and service transformers in cities and towns, so that if more than 2,500 volts are employed on the transmission line, step-down transformers are required at a sub-station. For a transmission of more than ten miles the saving in line conductors by operation at 10,000 to 12,000 volts will usually more than offset the additional cost of generators designed for this pressure and of step-down transformers. If the voltage of transmission is to exceed that of distribution, it will generally be found desirable to carry the former voltage up to 10,000 or 12,000, at least.

As the cost of generators designed for the voltage last named is less than that of lower voltage generators plus transformers, step-up transformers should usually be omitted in systems where these pressures are not exceeded. For alternating pressures above 13,000 to 15,000 volts, step-up transformers must generally be employed. In order that the saving in the weight of line conductors may more than offset the additional cost of transformers when the voltage of transmission is carried above 15,000, this voltage should be pushed on up to as much as 25,000 in most cases.

Power transmission with continuous current has the advantage that the cost of generators remains nearly the same whatever the line voltage, and that no transformers are required. Such transmissions are common in Europe, but have hardly a footing as yet in the United States. The reason for the uniform cost of continuous-current generators is found in the fact that they are connected in series to give the desired line voltage, and the voltage of each machine is kept under 3,000 or 4,000. As a partial offset to the low cost of the continuous-current generators and to the absence of transformers, there is the necessity for motor-generators in a sub-station when current for lighting as well as power is to be distributed.

In spite of the various additions to the cost of transmission systems made necessary by the adoption of very high voltages, these additions are much more than offset by the saving in the cost of conductors on lines 30, 50, or 100 miles in length. In fact, it is only by means of voltages ranging from 25,000 to 50,000 that the greatest of these distances, and others up to more than 140 miles, have been successfully covered by transmission lines. Above 60,000 volts there has been but slight practical experience in the operation of transmission lines.

Calculations to determine the sizes of conductors for electric transmission lines are all based on the fundamental law discovered by Ohm, which is that the electric current flowing in a circuit at any instant equals the electric pressure that causes the current divided by the electric resistance of the circuit itself, or current = pressure ÷ resistance.

Substituting in this formula the units that have been selected because of their convenient sizes for practical use, it becomes, amperes = volts ÷ ohms, in which the ohm is simply the electrical resistance, taken as unity, of a certain standard copper bar with fixed dimensions.

The ampere is the unit flow of current that is maintained with the unit pressure of one volt between the terminals of a one-ohm conductor. When this formula is applied to the computation of transmission lines the volts represent the electrical pressure that is required to force the desired amperes of current through the ohms of resistance in any particular line, and these volts have no necessary or fixed relation to the total voltage at which the line may operate. Thus, if the total voltage of a transmission system is 10,000, it may be desirable to use 500, 1,000, or even 2,000 volts to force current through the line, so that one of these numbers will represent the actual drop or loss of volts in the line conductors when the number of amperes that represent full load is flowing. As it is a law of every electric circuit that the rate of transformation of electric energy to heat or work in each of its several parts is directly proportional to the drop of voltage therein, it follows that a drop of 500 or 1,000 or 2,000 volts in the conductors of a 10,000-volt transmission line at full load would correspond to a power loss of five to ten or twenty per cent respectively. Any other part of 10,000 volts might be selected in this case as the pressure to be lost in the line. Evidently no formula can give the number of volts that should be lost in line conductors at full load for a given power transmission, but this number must be decided on by consideration of the items of line efficiency, regulation, and the ratio of the available power to the required load.

Having decided on the maximum loss of volts in the line conductors, and knowing the full voltage of operation, the power and consequently the number of amperes delivered to the line at maximum load, the resistance of the conductors may then be calculated by the formula, amperes = volts ÷ ohms. Thus, if the proposition is to deliver 2,000,000 watts or 2,000 kilowatts to a two-wire transmission line with a voltage of 20,000, the amperes in each wire must be represented by 2,000,000 ÷ 20,000 = 100. With a drop of ten per cent or 2,000 volts in the two line conductors, their combined resistance must be found from 100 = 2,000 ÷ ohms, and the ohms are therefore twenty. If the combined length of the two conductors is 200,000 feet, corresponding to a transmission line of a little under twenty miles, the resistance of these conductors must be 20 ÷ 200 = 0.1 ohm per 1,000 feet. From a wire table it may be seen that a No. 1/0 wire of copper, B. & S. gauge, with a diameter of 0.3249 inch, has a resistance of 0.1001 ohm per 1,000 feet at the temperature of 90° Fahrenheit, a little less at lower temperatures, and is thus the required size. Obviously, the resistance of twenty ohms is entirely independent of the length of the line, all the other factors being constant, and wires of various sizes will be required for other distances of transmission.

It is often convenient to find the area of cross section for the desired transmission conductor instead of finding its resistance. This can be done by substituting in the formula, amperes = volts ÷ ohms, the expression for the number of ohms in any conductor, and then solving as before.

Electrical resistance in every conductor varies directly with its length, inversely with its area of cross section, and also has a constant factor that depends on the material of which the conductor is composed. This constant factor is always the same for any given material, as pure iron, copper, or aluminum, and is usually taken as the resistance in ohms of a round wire one foot long and 0.001 inch in diameter, of the material to be used for conductors. Such a wire is said to have an area in cross section of one circular mil, because the square of its diameter taken as unity is still unity, that is, 1 × 1 = 1. In like manner, for the convenient designation of wires by their areas of cross-section, each round wire of any size is said to have an area in circular mils equal to the square of its diameter measured in units of 0.001 inch each. Thus, a round wire of 0.1 inch diameter has an area of 100 × 100 = 10,000 circular mils, and a round wire one inch in diameter has an area of 1,000 × 1000 = 1,000,000 circular mils. The circular mils of a wire do not express its area of cross section in terms of square inches, but this is not necessary since the resistance of a wire of one circular mil is taken as unity. Obviously, the areas of all round wires are to each other as are their circular mils.

From the foregoing it may be seen that the resistance of any round conductor is represented by the formula, ohms = $l \times s$ ÷ circular mils, in which l represents the length of the conductor in feet, s is the resistance in ohms of a wire of the same material as the conductor but with an area of one circular mil and a length of one foot, and the circular mils are those of the required conductor. Substituting the quantity, $l \times s \div$ circular mils, for ohms in the formula, amperes = volts ÷ ohms, the equation, amperes = volts ÷ ($l \times s \div$ circular mils), is obtained, and this reduces to circular mils = amperes $\times\ l \times s \div$ volts. For any proposed transmission all of the quantities in this formula are known, except the desired circular mils of the line conductors. The constant quantity s is about 10.8 for copper, but is conveniently used as eleven in calculation, and this allows a trifle for the effects of impurities that may exist in the line wire.

The case above mentioned, where 2,000 kilowatts were to be delivered to a transmission line at 20,000 volts, and a loss of 2,000 volts at full load was allowed in the line conductors, may now be solved by the formula for circular mils. Taking the resistance of a round copper wire 0.001 inch in diameter and one foot long as eleven ohms, and substituting the 100 amperes, 2,000 volts, and 200,000 feet of the present case in the formula, gives circular mils = (100 × 200,000 × 11) ÷ 2,000 = 110,000. The square root of this 110,000 will give the diameter of a copper wire that will exactly meet the conditions of the case, or the more practical course of consulting a table of standard sizes of wire will show that a No. 1-0 B. & S. gauge, with a diameter of 0.3249 inch, has a cross section of 105,500 circular mils, or about five per cent less than the calculated number, and is the size nearest to that wanted. As this No. 1-0 wire will give a line loss at full load of about 10.5 per cent, or only one-half of one per cent more than the loss at first selected, it should be adopted for the line in this case.

The formula just made use of is perfectly general in its application, and may be applied to the calculation of lines of aluminum or iron or any other metal just as well as to lines of copper. In order to use the formula for any desired metal, it is necessary that the resistance in ohms of a round wire of that metal one foot long and 0.001 inch in diameter be known and substituted for s in the formula. This resistance of a wire one foot long and 0.001 inch in diameter is called the specific resistance of the substance of which the wire is composed. For pure aluminum this specific resistance is about 17.7, for soft iron about sixty, and for hard steel about eighty ohms. The use of these values for s in the formula will therefore give the areas in circular mils for wires of these three substances, respectively, for any proposed transmission line. In the same way the specific resistance of any other metal or alloy, when known, may be applied in the formula.

The foregoing calculations apply accurately to all two-wire circuits that carry continuous currents, whether these circuits operate with constant current, constant pressure, or with pressure and current both variable. Where circuits are to carry alternating currents, certain other factors may require consideration. Almost all transmissions with alternating currents are carried out with three-phase three-wire, or two-phase four-wire, or single-phase two-wire circuits. Of the entire number of such transmissions, those with the three-phase three-wire circuits are in the majority, next in point of number come the two-phase transmissions, and lastly a few transmissions are carried out with single-phase currents. The voltage of a continuous-current circuit, by which the power of the transmission is computed and on which the percentage of line loss is based, is the maximum voltage operating there; but this is not true for circuits carrying alternating currents. Both the volts and amperes in an alternating circuit are constantly varying between maximum values in opposite directions along the wires. It follows from this fact that both the volts and amperes drop to zero as often as they rise to a maximum. It is fully demonstrated in books on the theory of alternating currents, that with certain ideal constructions in alternating generators, and certain conditions in the circuits to which they are connected, the equivalent or, as they are called, the virtual values of the constantly changing volts and amperes in these circuits are 0.707 of their respective maximum values. Or, to state the reverse of this proposition, the maximum volts and amperes respectively in these circuits rise to 1.414 times their equivalent or virtual values. These relations between maximum and virtual volts and amperes are subject to some variations with actual circuits and generators, but the virtual values of these factors, as measured by suitable volt- and amperemeters, are important in the design of transmission circuits, rather than their maximum values. When the volts or amperes of an alternating circuit are mentioned, the virtual values of these factors are usually meant unless some other value is specified. Thus, as commonly stated, the voltage of a single-phase circuit is the number of virtual volts between its two conductors, the voltage of a two-phase circuit is the number of virtual volts between each pair of its four conductors, and the voltage of a three-phase circuit is the number of virtual volts between either two of its three conductors.

Several factors not present with continuous currents tend to affect the losses in conductors where alternating currents are flowing, and the importance of such effects will be noted later. In spite of such effects, the formula above discussed should be applied to the calculation of transmission lines for alternating currents, and then the proper corrections of the results, if any are necessary, should be made. With this proviso as to corrections, the virtual volts and amperes of circuits carrying alternating currents may be used in the formula in the same way as the actual volts and amperes of continuous current circuits. Thus, reverting to the above example, where 2,000 kilowatts was to be delivered at 20,000 volts to a transmission line in which the loss was to be 2,000 volts, the kilowatts should be taken as the actual rate of work represented by the alternating current, and the volts named as the virtual volts on the line. The virtual amperes will now be 100, as were the actual amperes of continuous current, and the size of line conductor for a single-phase alternating transmission will therefore be 1-0, the same as for the continuous-current line. If the transmission is to be carried out on the two-phase four-wire system, the virtual amperes in each of these wires will be fifty instead of 100, as the power will be divided equally between the two pairs of conductors, and each of these four wires should have a cross-section in circular mils just one-half as great as that of the No. 1-0 wire. The required wire will thus be a No. 3 B. & S. gauge, of 52,630 circular mils, this being the nearest standard size. In weight the two No. 1-0 wires and the four No. 3 wires are almost equal, and they should be exactly equal to give the same loss in the single-phase and the two-phase lines. For a three-phase circuit to make the transmission above considered, each of the three conductors should have an area just one-half as great as that of each of the two conductors for a single phase circuit, the loss remaining as before, and the nearest standard size of wire is again No. 3, as it was for the two-phase line. This is not a self-evident proposition, but the proof can be found in books devoted to the theory of the subject. From the foregoing it is evident that while the single-phase and two-phase lines require equal weights of conductors, all other factors being the same, the weight of conductors in the three-phase line is only seventy-five per cent of that in either of the other two. Neglecting the special factors that tend to raise the size and weight of alternating-current circuits, the single-phase and two-phase lines require the same weight of conductors as does a continuous-current transmission of equal power, voltage, and line loss. It should be noted that in each of these cases the factor l in the formula for circular mils denotes the entire length of the pair of conductors for a continuous-current line, or double the distance of the transmission with either of the alternating-current lines.

Having found the circular mils of any desired conductor, its weight per 1,000 feet can be found readily in a wire table. In some cases it is desirable to calculate the weight of the conductors for a transmission line without finding the circular mils of each, and this can be done by a modification of the above formula. A copper wire of 1,000,000 circular mils weighs nearly 3.03 pounds per foot of its length, and the weight of any copper wire may therefore be found from the formula, pounds = (circular mils × 3.03 × l) ÷ 1,000,000, in which pounds indicates the total weight of the conductor, l, its total length, and the circular mils are those of its cross-section. This formula reduces to the form, circular mils = (1,000,000 × pounds) ÷ (3.03 × l) and if this value for circular mils is substituted in the formula above given for the cross-section of any wire, the result is (1,000,000 × pounds) ÷ (3.03 × l) = (l × amperes × 11) ÷ volts. Transposition of the factors in this last equation brings it to the form, pounds = (3.03 × l^2 × amperes × 11) ÷ (1,000,000 × volts), which is the general formula for the total weight of copper conductors when l, the length of one pair, the total amperes flowing, and the volts lost in the conductors are known for either a continuous-current, a single-phase, or a two-phase four-wire line.

If the value of *l*, 200,000, of amperes, 100, and of volts, 2,000, for the transmission above considered are substituted in the formula for total weight, just found, the result is pounds = $(3.03\ (200{,}000)^2 \times 100 \times 11) \div (1{,}000{,}000 \times 2{,}000)$, which reduced to pounds = 66,660, the weight of copper wire necessary for the transmission with either continuous, single-phase or two-phase current. With three-phase current the weight of copper in the line for this transmission will be 75 per cent of the 66,660 pounds just found. One or more two-wire circuits may be employed for the continuous current or for the single-phase transmission, and if one such circuit is used the weight for each of the two wires is obviously 33,660 pounds. For a two-phase transmission two or more circuits of two wires each will be used, and in the case of two circuits, if all four of the wires are of equal cross section, as would usually be the case, the total weight of each is 16,830 pounds. If the transmission is made with one three-phase circuit, the weight of each of the three wires is 16,830 pounds, and their combined weight, 50,490 pounds of copper. In each of these transmission lines the length of a single conductor in one direction is 100,000 feet, or one-half of the length of the wires in a single two-wire circuit. For the two-wire line the calculated weight of each conductor amounts to 66,660 ÷ 200 = 333.3 pounds per 1,000 feet. For a two-phase four-wire line and also for a three-phase three-wire line, the weight of each conductor is 16,830 ÷ 100 = 168.3 pounds per 1,000 feet. On inspection of a table of weights for bare copper wires it may be seen that a No. 1-0 B. & S. gauge wire has a weight of 320 pounds per 1,000 feet, and being much the nearest size to the calculated weight of 333 pounds should be selected for the two-wire circuit. It may also be seen that a No. 3 wire, with a weight of 159 pounds per 1,000 feet, is the size that comes nearest to the calculated weight of 168 pounds, and should therefore be employed in the three-wire and the four-wire circuits, for two- and three-phase transmissions. Either a continuous-current, single-phase, two-phase, or three-phase transmission line may of course be split up into as many circuits as desired, and these circuits may or may not be designed to carry equal portions of the entire power. In either case the combined weights of the several circuits should equal those above found, the conditions as to power, loss, and length of line remaining constant. It will be noted that the formulæ for the calculation of the circular mils and for the weight of the conductors in the transmission line lead to the selection of the same sizes of wires, as they obviously should do.

Several laws governing the relations of volts lost, length and weight of line conductors, may be readily deduced from the above formulæ. Evidently the circular mils and weight of line conductors vary inversely with the number of volts lost in them when carrying a given current, so that doubling this number of volts reduces the circular mils and weight of conductors by one-half. If the length of the line changes, the circular mils of the required conductors change directly with it, but the weight of these conductors varies as the square of their length. Thus, if the length of the line conductors is doubled, the cross-section in circular mils of each conductor is also doubled, and each conductor is therefore four times as heavy as before for the same current and loss in volts. Should the length of the conductors and also the number of volts lost in them be varied at the same rate, the circular mils of each conductor remain constant, and its weight increases directly with the distance of transmission. Thus, with the same size of line wire, both the number of volts lost and the total weight are twice as great for a 100- as for a fifty-mile transmission. If the total weight of conductors is to be held constant, then the number of volts lost therein must vary as the square of their length, and their circular mils must vary inversely as the length. So that if the length of a transmission line is doubled, the circular mils for conductors of constant weight are divided by two, and the volts lost are four times as great as before. Each of these rules assumes that the watts and percentage of loss in the line are constant.

The above principles and formulæ apply to the design of transmission lines for either continuous or alternating currents, but where the alternating current is employed certain additional factors should be considered. One of these factors is inductance, by which is meant the counter-electromotive force that is always present and opposed to the regular voltage in an alternating current circuit. One effect of inductance is to cut down the voltage at that end of the line where the power is delivered to a sub-station, just as is also done by the ohmic resistance of the line conductors. Between the loss of voltage due to line resistance and the loss due to inductance there is the very important difference that the former represents an actual conversion of electrical energy into heat, while the latter is simply the loss of pressure without any material decrease in the amount of energy. While the loss of energy in a transmission line depends directly on its resistance, the loss of pressure due to inductance depends on the diameter of conductors without regard to their resistance, on the length of the circuit, the distance between the conductors, and on the frequency or number of cycles per second through which the current passes. As a result of these facts, it is not desirable or even practicable to use inductance as a factor in the calculation of the resistance or weight of a transmission line. On transmission lines, as ordinarily constructed, the loss of voltage due to inductance generally ranges between 25 and 100 per cent of the number of volts lost at full load because of the resistance of the conductors. This loss through inductance may be lowered by reducing the diameter of individual wires, though the resistance of all the circuits concerned in the transmission remains the same, by bringing the wires nearer together and by adopting smaller frequencies. In practice the volts lost through inductance are compensated for by operating generators or transformers in the power-plant at a voltage that insures the delivery of energy in the receiving-station at the required pressure. Thus, in a certain case, it may be desirable to transmit energy with a maximum loss of ten per cent in the line at full load, due to the resistance of the conductors, when the effective voltage at the generator end of the line is 10,000, so that the pressure at the receiving-station will be 9,000 volts. If it appears that the loss of pressure due to inductance on this line will be 1,000 volts, then the generators should be operated at 11,000 volts, which will provide for the loss of 1,000 volts by inductance, leave an effective voltage of 10,000 on the line, and allow the delivery of energy at the sub-station with a pressure of 9,000 volts, when there is a ten-percent loss of power due to the line resistance.

Inductance not only sets up a counter-electromotive force in the line, which reduces the voltage delivered to it by generators or transformers, but also causes a larger current to flow in the line than is indicated by the division of the number of watts delivered to it by the virtual voltage of delivery. The amount of current increase depends on both the inductance of the line itself and also on the character of its connected apparatus. In a system with a mixed load of lamps and motors there is quite certain to be some inductance, but it is very hard to predetermine its exact amount. Experience with such systems shows, however, that the increase of line current due to inductance is often not above five and usually less than ten per cent of the current that would flow if there were no inductance. To provide for the flow of this additional current, due to inductance, without an increase of the loss in volts because of ohmic resistance, the cross section of the line conductors must be enlarged by a percentage equal to that of the additional current. This means that in an ordinary case of a transmission with either single, two, or three-phase alternating current, the circular mils of each line wire, as computed with the formulæ above given, should be increased by five to ten per cent. Such increase in the cross section of wires of course carries with it a like rise in the total weight of the conductors for the transmission. If wire of the cross section computed with the formulæ is employed for the alternating current transmission, inductance in an ordinary case will raise the assumed line loss of power by five to ten per cent of what it would be if no inductance existed. Thus, with conductors calculated by the formulæ for a power loss of ten per cent at full load, inductance in an ordinary case would raise this loss to somewhere between 10.5 and eleven per cent. As a rule it may therefore be said that inductance will seldom increase the weight of line conductors, or the loss of power therein, by more than ten per cent.

When an alternating current flows along a conductor its density is not uniform in all parts of each cross section, but the current density is least at the centre of the conductor and increases toward the outside surface. This unequal distribution of the alternating current over each cross section of a conductor through which it is passing increases with the diameter or thickness of the conductor and with the frequency of the alternating current. By reason of this action the ohmic resistance of any conductor is somewhat greater for an alternating than for a continuous current, because the full cross section of the conductor cannot be utilized with the former current. Fortunately, the practical importance of this unequal distribution of alternating current over each cross section of its conductor is usually slight, so far as the sizes of wires for transmission lines are concerned, because the usual frequencies of current and diameters of conductors concerned are not great enough to give the effect mentioned a large numerical value. Thus, sixty cycles per second is the highest frequency commonly employed for the current on transmission lines. With a 4-0 wire, and the current frequency named, the increase in the ohmic resistance for alternating over that for continuous current does not reach one-half of one per cent.

Having calculated the circular mils of weight of a transmission line by the foregoing formulæ, it appears that the only material increase of this weight required by the use of alternating current is that due to inductance. This increase cannot be calculated exactly beforehand because of the uncertain elements in future loads, but experience shows that it is seldom more than ten per cent of the calculated size or weight of conductors.

CHAPTER XVII.
SELECTION OF TRANSMISSION CIRCUITS.

Maximum power, voltage, loss, and weight of conductors having been fixed for a transmission line, the number of circuits that shall make up the line, and the relations of these circuits to each other, remain to be determined.

In practice wide differences exist as to the number and relations of circuits on a single transmission line between two points. Cases illustrating this fact are the 147-mile transmission from Electra power-house to San Francisco and the 65-mile transmission between Cañon Ferry, on the Missouri River and Butte, Mont. At the Electra plant the generator capacity is 10,000 kilowatts, and the transmission to San Francisco is carried out over a single pole line that carries one circuit composed of three aluminum conductors, each with an area in cross section of 471,000 circular mils. From the generators at Cañon Ferry, which have an aggregate capacity of 7,500 kilowatts, a part of the energy goes to Helena over a separate line, and the transmission to Butte goes over two pole lines that are 40 feet apart. Each of these two pole lines carries a single circuit composed of three copper conductors, and each conductor has a cross section of 105,600 circular mils. The difference in practice illustrated by these two plants is further brought out by the fact that their voltages are not far apart, as the Cañon Ferry and Butte line operates at 50,000, and the Electra and San Francisco line at 60,000 volts.

Economy in the construction of a transmission line points strongly to the use of a single circuit, because this means only one line of poles, usually but one cross-arm for the power wires per pole, the least possible number of pins and insulators, and the smallest amount of labor for the erection of the conductors. In favor of a single circuit there is also the argument of greatest mechanical strength in each conductor, since the single circuit is to have the same weight as that of all the circuits that may be adopted in its place. Where each conductor of the single circuit would have a cross section of less than 83,690 circular mils, if of copper, corresponding to a No. 1 B. & S. gauge wire, the argument as to mechanical strength is of especial force, since two equal circuits instead of one, in the case where one circuit of No. 1 wires would have the required weight, reduce the size of each conductor to No. 4 wire, of 41,740 circular mils cross section, and this is the smallest wire that it is practicable to use on long lines for mechanical reasons. Opposed to these arguments for a single circuit are those based on the supposed greater reliability of two or more circuits, their greater ease of repair, their more effective means of regulation, and the influence on inductance of a reduction in the size of conductors.

In spite of the consequent reduction in the size of each conductor, the use of two or more separate circuits for the same transmission is sometimes thought to increase its reliability, because in case of a break or short-circuit on one of the circuits the other will still be available. Breaks in transmission conductors are due either to mechanical strains alone, as wind pressure, the falling of trees, or the accumulation of ice, or else to an arc between the conductors that tends to melt them at some point. As a smaller conductor breaks or melts more readily than a large one, the use of two or more circuits instead of a single circuit tends to increase troubles of this sort. It thus seems that while two or more circuits give a greater chance of continued operation after a break in a conductor actually occurs, the use of a single circuit with larger conductors makes any break less probable.

When repairs must be made on a transmission line, as in replacing a broken insulator or setting a pole in the place of one that has burned, it is certainly convenient to have two or more circuits so that one may be out of use while the repairs on it are made. It is practicable, however, to make such repairs on any high-voltage circuit, even when it is in use, provided the conductors are spaced so far apart that there is no chance of making a contact or starting an arc between them. To get such distance between conductors there should be only one circuit per pole, and even then more room should be provided for that circuit than is common in this type of construction. On each of the two pole lines between Cañon Ferry and Butte there is a single circuit of three conductors arranged in triangular form, two at the opposite ends of a cross-arm and one at the top of the pole, and the distance from each conductor of a circuit to either of the other two is 6.5 feet. This distance between conductors is perhaps as great as that on any transmission circuit now in use, but it seems too small to make repairs on the circuit reasonably safe when it is in operation at a pressure of 50,000 volts. There seems to be no good reason why the distance between the conductors of a single circuit to which a pole line is devoted might not be increased to as much as ten feet, at the slightly greater expense of longer cross-arms. With as much as ten feet between conductors, and special tools with long wooden handles to grasp these conductors, there should be no serious danger about the repair of even 60,000-volt lines when in operation. As the 60,000-volt line between Electra and San Francisco consists of only one circuit, it seems that repairs on it must be contemplated during operation.

Another example of a high-voltage transmission carried out with a single circuit is that between Shawinigan Falls and Montreal, a distance of eighty-five miles. In this case the circuit is made up of three aluminum conductors, each of which has an area in cross section of 183,750 circular mils, and these conductors are located five feet apart, one at the top of each pole, and two at the ends of a cross-arm below. This single circuit is in regular operation at 50,000 volts for the supply of light and power in Montreal, and it is hard to see how repairs while there is current on the line are to be avoided.

Inductance varies with the ratio between the diameter of the wires in any circuit and the distance between these wires, but as inductance simply raises the voltage that must be delivered by generators or transformers, and does not represent a loss of energy, it may generally be given but little weight in selecting the number of circuits, the distance between conductors, and the size of each conductor. If two or more circuits with smaller conductors have a combined resistance in multiple equal to that of a single circuit with larger conductors, the loss of voltage due to inductance may be greater on the single circuit than the corresponding loss on the multiple circuits, but the advantages due to the single circuit may more than compensate for the higher pressure at generators or transformers. That such advantages have been thought to exist in actual construction may be seen from the fact that the 147-mile line from Electra power-house to San Francisco, and the 83-mile line from Shawinigan Falls to Montreal, are composed of one circuit each. As inductance increases directly with the length of circuits, these very long lines are especially subject to its influence, yet it was thought that the advantages of a single circuit more than offset its disadvantages in each case.

Where several sub-stations, widely separated, are to be supplied with energy by the same transmission line, another argument exists for the division of the line conductors into more than one circuit, so that there may be an independent circuit to each sub-station. As the pressure for local distribution lines must be regulated at each sub-station, it is quite an advantage to have a separate transmission circuit between each sub-station and the power plant, so that the voltage on each circuit at the power-house may be adjusted as nearly as possible to the requirements of its sub-station. An interesting illustration of this practice may be noted in the design of transmission circuits for the line between Spier Falls on the Hudson River and the cities of Schenectady, Troy, and Albany, located between thirty and forty miles to the south, which passes through Saratoga and Ballston on the way. When this transmission line is completed, four three-phase circuits, one of No. 0 and three of No. 000 copper wire, will run to the Saratoga switch-house from the generating plant at the Falls, a distance of some eight miles.

From this switch-house two circuits of No. 0 conductors go to the Saratoga sub-station, a little more than one mile away, two circuits of No. 000 wires run to the Watervliet sub-station, across the river from Troy and thirty-five miles from the generating station, and one circuit of No. 0 and one circuit of No. 000 wires are carried to Schenectady, thirty miles from Spier Falls, passing through and supplying the Ballston sub-station on the way. Other circuits connect the sub-station at Watervliet with that at Schenectady and with the water-power station at Mechanicsville. From the Watervliet sub-station secondary lines run to sub-stations that control the local distribution of light and power in Albany and Troy. This network of transmission circuits was made desirable by the conditions of this case, which include the general supply of light and power in three large and several smaller cities, the operation of three large electric railway systems, and the delivery of thousands of horse-power for the motors in a great manufacturing plant.

In not every transmission system with different and widely scattered loads it is thought desirable to provide more than one main circuit. Thus, the single circuit eighty-three miles long that transmits energy from Shawinigan Falls to Montreal is designed to supply power also in some smaller places on the way.

So again, the 147-mile circuit from Electra power-house to San Francisco passes through a dozen or more smaller places, including Stockton, and is tapped with side lines that run to Oakland and San José. In cases like this, where very long lines run through large numbers of cities and towns that sooner or later require service, it is obviously impracticable to provide a separate circuit for each centre of local distribution. It may well be in such a case that a single main transmission circuit connected to a long line of sub-stations will represent the best possible solution of the problem. At the power-house end of such a circuit the voltage will naturally be regulated to suit that sub-station where the load is the most important or exacting, and each of the other sub-stations will be left to do all of the regulating for its own load.

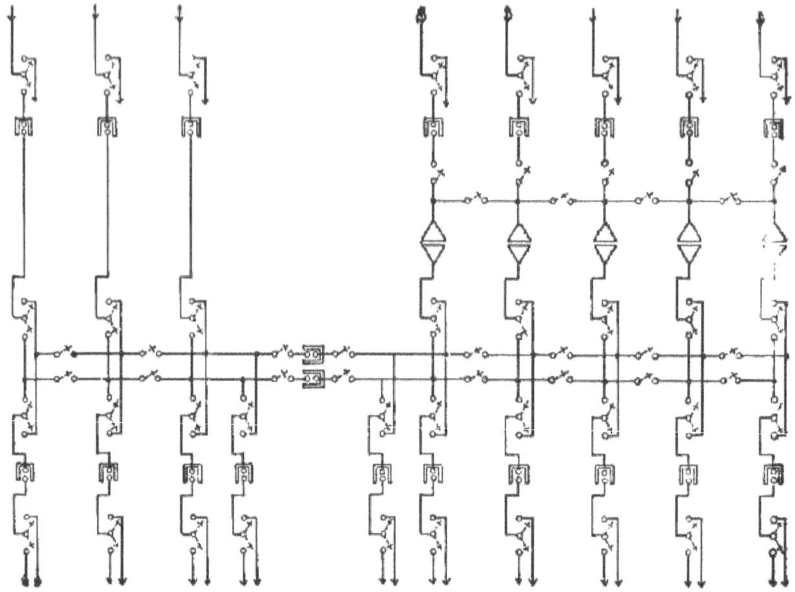

FIG. 76.—Connections at Watervliet Sub-station on Spier Falls Line.

The greater the total loss of voltage on a transmission line supplying sub-stations that are scattered along much of its length, the larger will be the fluctuations of voltage that must be compensated for at all of the sub-stations save one, under changing loads, if only one circuit is employed between the power-plant and these sub-stations. Suppose, for example, that a transmission line 100 miles long is composed of a single circuit, and supplies two sub-stations, one located 50 miles and the other 100 miles from the power-plant. Assume at first that there is no load whatever at the intermediate sub-station. If the single transmission circuit operates with 50,000 volts at the power-plant, and 45,000 volts at the sub-station 100 miles away when there is a full load there, corresponding to a loss of ten per cent, then the pressure at the intermediate sub-station will be 47,500 volts. If, now, the load at the sub-station 100 miles from the power-house drops to a point where the entire line loss is only 1,000 volts, and the pressure at the generating plant is lowered to 46,000 volts so as to maintain 45,000 volts at the more distant sub-station, then the pressure at the intermediate sub-station will be 45,500 volts, or 2,000 volts less than it was before. If the loss on the entire line at full load were only five per cent, making the voltage at the sub-station 100 miles away 47,500 when that at the generating station is 50,000, then the pressure at the intermediate sub-station will be 48,750 volts. Upon a reduction of the loss on the entire length of line to one-fifth of its maximum amount, or to 500 volts, the pressure at the generating station must be reduced to 48,000 volts, if that at the more distant sub-station is to be held constant at 47,500. At the intermediate sub-station the pressure will then be 47,750 volts, or 1,000 volts less than it was at full load. From these two examples it may be seen that the extent of pressure variation at the intermediate sub-station will depend directly on the maximum line loss, if the regulation at the generating station is such as to maintain a constant voltage at the sub-station 100 miles away.

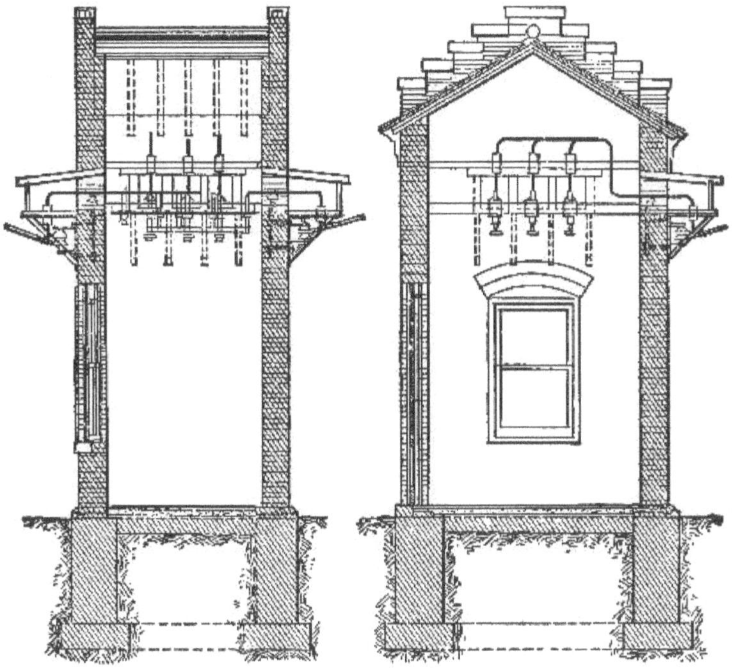

FIG. 77.—Sections of Switch-house on New Hampshire Traction System.

All the foregoing has assumed no load to be connected at the intermediate sub-station, and with a load there the fluctuations of pressure will of course depend on its amount as well as on the load at the more distant sub-station.

One of the strongest reasons for the use of two or more circuits in the same transmission line arises from the rapid fluctuations of load where large stationary motors or an electric railway system is operated. When a transmission line must carry a load of stationary or railway motors, it is a common practice to divide the line into at least two circuits, and to devote one circuit exclusively to railway or motor work and another to lighting, at any one time. In some cases this division of the transmission system into two parts, one devoted to the lighting and the other to the motor load, is carried out not only as to the sub-station apparatus and the line, but also as to the transformers, generators, water-wheels, and even the penstocks at the power-plant. It is possible even to carry this division of the transmission system still further, and to separate either the motor or the lighting load, or both, into sections, and then to devote a distinct transmission circuit, group of transformers, generator, and water-wheel to the operation of each section. An example of the complete division of generating and transmitting apparatus into independent units may be noted in the case of the system that supplies light and power in Portland, Me., from a generating plant on the Presumpscot River, thirteen miles away. At this station four steel penstocks, each provided with a separate gate at the forebay wall, bring water to as many pairs of wheels, and each pair of wheels drives a direct-connected generator. Four three-phase circuits connect the generating plant with the sub-station at Portland, and each circuit between the generating plant and a transformer-house outside the business section of the city is made up of No. 2 solid soft-drawn copper wires.

Each of these four sets of apparatus, from head-gate to sub-station, is usually operated independently of the others, and supplies either the motor load or a part of the electric lighting. In this way changes in the amount of one section of the load cause no fluctuation of the voltage on the other sections. At Manchester, N. H., the sub-station receives energy from four water-power plants, and is provided with two sets of low-tension, 2,300-volt, three-phase bus-bars, one set of these bus-bars being devoted to the operation of the local electric railway system, and the other set to the supply of lamps and stationary motors. Each set of these bus-bars is divided into a number of sections, and by means of these sections different transmission circuits are devoted to different portions of the lighting and motor loads. As three of the four water-power plants are connected to the sub-station by two circuits each, the division of loads in this case is often carried clear back to the generators, one generator in a power-house being operated, for instance, on railway work and another on a lighting load at the same time. This plan has the obvious advantage that much of the regulation for the several parts of the entire load may be done at the generators, thus reducing the amount of regulation necessary at the sub-station, and that fluctuating motor loads do not affect the lamps. In this case the conductors of the several transmission circuits are all of moderate size, and the division of the lines was evidently adopted for purposes of regulation, rather than to reduce the amount of inductance. Thus the line between Gregg's Falls and the sub-station, a distance of six miles, is made up of one three-phase circuit of No. 4 and one circuit of No. 6 bare copper wires. The fourteen-mile line between the plant at Garvin's Falls and the sub-station, the longest of the four transmissions, is made up of two three-phase circuits, each composed of No. 0 bare copper wires. In the case of the Gregg's Falls plant the subdivision of the line has gone further than that of the generating equipment, for the station there contains only a single generator, the rating being 1,200 kilowatts, while two circuits run thence to the sub-station. Another instance showing extensive subdivision of a line into separate circuits may be noted in the seven-mile transmission from Montmorency Falls to Quebec, Canada, where sixteen conductors, each No. 0 copper wire, make up four two-phase circuits that connect a plant of 2,400 kilowatts capacity with its sub-station.

Such multiplication of transmission circuits has some advantages from the standpoint of regulation, but there are good reasons for limiting it to rather short lines, where it is, in fact, almost exclusively found. On very long lines the use of numerous circuits composed of rather small conductors would obviously increase the constant expense of inspection and repairs and add materially to uncertainty of the service. Very few, if any, transmission lines of as much as twenty-five miles in length are divided into more than two circuits, and in several instances lines of superlative length have only a single circuit each. The greatest single power transmission in the world, that between Niagara Falls and Buffalo, is carried out with two pole lines, one of which is about twenty and the other about twenty-three miles long. The longer pole line, which is also the older, carries two three-phase circuits, each of which is made up of three 350,000 circular mil copper conductors. The shorter pole line carries a single three-phase circuit composed of aluminum conductors, each of which has an area in cross section of 500,000 circular mils. In electrical conductivity the aluminum circuit is intended to be equal to each of the two that are composed of copper. According to the description of the Niagara Falls and Buffalo transmission system in vol. xviii., A. I. E. E., pages 518 to 527, each of these three circuits is designed to transmit about 7,500 kilowatts, and the maximum power transmitted up to August, 1901, was 15,600 kilowatts, or about the calculated capacity of two of the circuits. According to the description just mentioned, the transmission circuits used to supply energy for use at Buffalo are regularly operated in parallel, and this is also true of the generators and the step-down transformers, though the uses to which this energy is applied include lighting, large stationary motors, and the electric railway system. Apparatus in the generating station at Niagara Falls and in the terminal-house near the city limits of Buffalo is so arranged, however, that two of the 3,750 kilowatt generators and eight step-up transformers at the power-house, together with one transmission circuit and three step-down transformers in the terminal-house at Buffalo, may be operated independently of all the other apparatus.

As already pointed out, the use of separate circuits for each sub-station, and for lighting and power loads at each sub-station in very long transmission systems, is often impracticable. Even in comparatively short transmissions the multiplication of circuits and the use of rather small and mechanically weak conductors increased the first cost of installation and the subsequent expense of inspection and repairs. An objection to operation with a single circuit in a transmission line that supplies widely separated sub-stations with lighting, power, and railway loads is the consequent difficulty of pressure regulation on the distribution lines at each sub-station. Such a transmission line necessarily delivers energy at different and fluctuating voltages at the several sub-stations, and these fluctuations are of course reproduced on the secondary side of the step-down transformers. Fortunately, however, the use of synchronous motor generators, either in place of or in connection with static transformers, goes far to solve the problem of pressure regulation for distribution circuits supplied with energy from transmission lines. This is due to the well-known fact that with constant frequency the speed of rotation for a synchronous motor is constant without regard to fluctuations in the applied voltage or changes in its load. With a constant speed at the motor and its connected generator it is of course easy to deliver current at constant voltage to the distribution lines. This constancy of speed makes the synchronous motor generator a favorite in large transmission systems with both power and lighting loads. The satisfactory lighting service in Buffalo, operated with energy transmitted from Niagara Falls, seems to be due in some measure to the use of synchronous motor generators at the sub-station in Buffalo, whence lighting circuits are supplied. As above stated, the three circuits that make up the transmission line between Niagara Falls and Buffalo are operated in multiple, and in the latter place there is a large load of both railway and stationary motors. As the three circuits are operated in multiple, they of course amount to only a single circuit so far as fluctuations of voltage due to changes in these several sorts of loads are concerned. According to vol. xviii., A. I. E. E., pages 125 and following, the load on the transmission system at Buffalo in 1901 was made up of about 7,000 horse-power in railway motors, 4,000 horse-power in induction motors, and 4,000 horse-power divided up between series arc lamps, constant pressure incandescent lamps, and continuous current motors. The railway load is operated through step-down transformers and rotary converters. The induction motors are connected either to the 2,000-volt secondary circuits of the step-down transformers or to service transformers supplied by these circuits. On these railway and stationary motor loads there is of course no necessity for close pressure regulation. Series arc lamps are operated through step-down transformers and synchronous motors direct-connected to constant continuous current dynamos. Continuous current stationary motors draw power from the transmission lines through step-down transformers and rotary converters, like the railway load. For the 2,200 volt circuits that supply service transformers for commercial arc and incandescent lighting the transmitted energy passes through step-down transformers and synchronous motor-generators. These motor-generators raise the frequency from twenty-five to sixty cycles per second. Finally the continuous current three-wire system for incandescent lighting at about 250 volts between outside wires is operated through step-down transformers and synchronous motors direct-connected to continuous current generators. For this last-named service rotary converters were at first tried, but were found to be impracticable because voltage fluctuations on the transmission line (due largely to the railway and motor loads) were reproduced on the continuous-current circuits by the rotary converters. Since the adoption of motor-generators this fluctuation of the service voltage is no longer present.

Another case in which synchronous motor-generators deliver power from a transmission line that carries both a lighting and a motor load is that of the Shawinigan sub-station in Montreal. At this sub-station the 85-mile transmission line from the generating plant at Shawinigan Falls terminates. As already pointed out, this line is composed of a single three-phase circuit of aluminum conductors, each of which has a cross section of 183,750 circular mils. In the Montreal sub-station the thirty-cycle, three-phase current from Shawinigan Falls is delivered to transformers that lower the voltage to 2,300. The current then goes to five synchronous motor-generators of 1,200 horse-power capacity each, and is there converted to sixty-three cycles per second, two-phase, at the same voltage. This converted current passes onto the distribution lines of the local electrical supply system in Montreal, which also draws energy from two other water-power plants, and is devoted to lighting, stationary motors, or to the street railway work, as may be required. Though separate local distribution circuits are devoted to these several loads, the fluctuations in the stationary and railway motor work necessarily react on the voltage of the transmission line and transformers at the sub-station. By the use of the synchronous motor-generators the lighting circuits are protected from these pressure variations.

As the numbers of sub-stations at different points on long transmission lines increase, and stationary motor and railway loads at each become more common, it is to be expected that the use of synchronous motor-generators for lighting service will be much more frequent than at present. With such use there will disappear one of the reasons for the multiplication of transmission circuits.

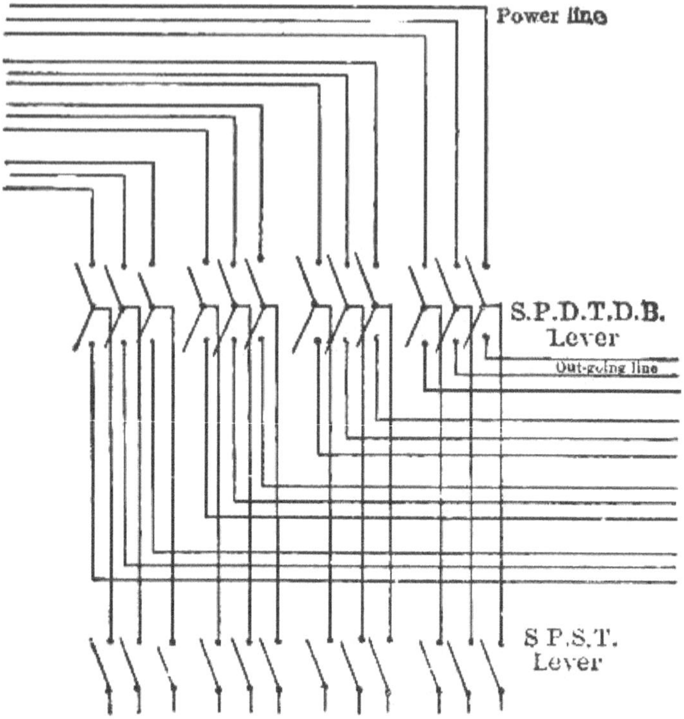

FIG. 78.—Transfer Switches at Saratoga Switch-house on Spier Falls Line.

Where several transmission circuits connect a generating plant with a single sub-station, or with several sub-stations in the same general direction, it is desirable to have switches so arranged that two or more circuits may be combined as one, or so that any circuit that ordinarily operates a certain load or sub-station may be devoted to another when occasion requires. For this purpose transfer switches on each circuit are necessary at generating plants, sub-stations, and often at switch-houses. These transfer switches will ordinarily be of the knife type, and intended for manual operation when the circuits to which they are connected are not in use. As such switches are exposed to the full voltage of transmission, the insulation of their conducting parts should be very high. In the extensive transmission system between the power-plants at Spier Falls and Mechanicsville and the sub-stations at Troy, Albany, and Schenectady, N. Y., a transfer switch of highly insulated construction has been much used. The two blades of this switch move independently of each other, but both are mounted between the same metal clips. Each blade is of two by one-quarter inch drawn copper rod, and the clips supporting the two blades are mounted on top of a circular metal cap four and three-quarter inches in outside diameter and two inches high, that is cemented over the top of a large, double petticoat, porcelain line insulator.

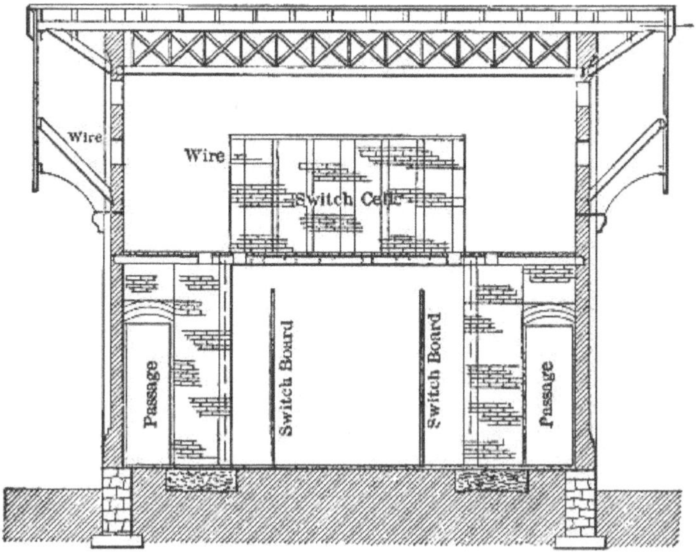

FIG. 79.—Cross Section of Schenectady Switch-house on Spier Falls Line.

Clips into which these copper blades are swung in closing the switch are also mounted in caps carried by insulators in the way just described. Each of these insulators is mounted on a large wooden pin, and these pins are secured in timbers at the points where the switches are wanted. This construction of switches gives ample insulation for the line voltage of 30,000 in this system. By means of the transfer switches just described, either of the transmission circuits leaving the Spier Falls power-plant may be connected to any one of the ten generators and ten groups of transformers there. At the Saratoga switch-house, any one of the twelve conductors, making up the four three-phase circuits from Spier Falls may be connected to any one of the eighteen conductors making up the six three-phase circuits that go south to Saratoga, Watervliet, and Schenectady sub-stations, in the way indicated by the drawing. So again at the Watervliet sub-station, where energy at 26,500 volts is received from Spier Falls and energy at 10,800 volts from Mechanicsville, any single conductor from either of these water-power plants may be connected, either directly or through a transformer, with either conductor running to the railway and lighting sub-stations about Albany and Troy. Where several transmission circuits are employed, this complete flexibility of connection evidently adds materially to the convenience and reliability of operation.

CIRCUITS IN TRANSMISSION LINES.

Location of Lines.	Length in Miles.	Number of Circuits.	Number of Pole Lines.	Circular Mils per Wire.	Cycles per Second of Current.
Electra to San Francisco	147	1	1	[A]471,034	60
Colgate to Oakland, Cal.	142	2	2	133,100	60
				[A]211,000	
Santa Ana River to Los Angeles	83	2	1	83,690	60
Shawinigan Falls to Montreal	85	1	1	[A]183,750	30
Cañon Ferry to Butte	65	2	2	106,500	60
Welland Canal to Hamilton	35	1	1	83,690	60
Welland Canal to Hamilton	37	1	1	133,100	60
Spier Falls to Schenectady	30	2	1	105,600	40
				167,800	
Spier Falls to Watervliet, N. Y.	35	2	1	167,800	40
Ogden to Salt Lake City	36	2	1	83,690	60
Apple River Falls to St. Paul	27	2	1	66,370	60
Niagara Falls to Buffalo	23	2	1	350,000	25
Niagara Falls to Buffalo	20	1	1	[A]500,000	25
Farmington River to Hartford	11	1	1	[A]364,420	60
Niagara Falls to Toronto	75	2	1 [B]	190,000	25

[A] Aluminum conductor.

[B] Steel towers.

CHAPTER XVIII.
POLE LINES FOR POWER TRANSMISSION.

Long transmission lines should follow the most direct routes between generating and sub-stations as far as practicable. The number of poles, cross-arms, and insulators increases directly with the length of line, and the weight of conductors with the square of that length, other factors remaining equal. Every material deviation from a straight line must therefore be paid for at a rather high rate.

Distribution lines necessarily follow the public streets in order to reach consumers, but the saving of the cost of a private right of way and ease of access are the main considerations which tend to keep transmission lines on streets and highways. Except in very rough or swampy country, the difficulty of access to a pole line on a private right of way is not a serious matter and should be given but little weight. The cost of a private right of way may be more important, and should be compared with the additional cost of the pole line and conductors if erected on the public highway. In this additional cost should be included any items for paving about the poles, extra pins, insulators, and guys made necessary by frequent turns in the highway, and the sums that may be required to secure the necessary franchises. There is also the possible contingency of future legislation as to the voltage that may be maintained on wires located over public streets. These considerations taken together give a strong tendency to the location of long transmission lines on private rights of way, especially where the amount of power involved is great and the voltage very high.

A transmission line 80.3 miles in length recently erected between Rochester and Pelham, N. H., by way of Portsmouth, where the generating station is located, to feed an electric railway system, operates at 13,200 volts and is mainly located on private rights of way. Deeds conveying the easements for this right of way provide that all trees or branches within one rod on either side of the line may be cut away. The transmission line between Niagara Falls and Buffalo, about twenty-three miles long and operating at 22,000 volts, is largely on a private way thirty feet wide.

FIG. 80.—Transmission Line of New Hampshire Traction Company over Hampton River Bridge, 4,623 Feet Long.

For the transmission between Cañon Ferry and Butte the line is mainly located on a private way. Between Colgate and Oakland the transmission line is mostly on private way, and this is also true of the greater part of some other high-pressure lines in California. These private rights of way range from fifty to several hundred feet wide, it being necessary in forests to cut down all trees that are tall enough to fall onto the wires.

In some cases of transmission at very high voltage two independent pole lines are erected and one or more circuits are then run on each set of poles. This construction has been followed on the transmission line between Niagara Falls and Buffalo, Cañon Ferry and Butte, Welland Canal and Hamilton, and between Colgate and Oakland. Such double pole lines are more usually located on the same right of way, this being true of the Cañon Ferry and Colgate systems, but this is not always the case. In the Hamilton system the two lines of poles, one thirty-five miles and the other thirty-seven miles in length, are located several miles apart. The two sets of poles on a part of the Buffalo line are less than thirty feet, on the Colgate line are twenty-five feet, and on the Cañon Ferry line are forty feet apart.

The main reasons for the use of two pole lines instead of one are the probability that an arc started on one circuit will be communicated to another on the same poles, and the greater ease and safety of repairs when each circuit is on a separate line of poles. On each pole line of the Cañon Ferry transmission, and also on each pole line of the Colgate transmission, there is only one three-wire circuit. On the Cañon Ferry line each wire of the two circuits has a cross-section of only 106,500 circular mils, and on the Colgate line one circuit is of 133,225 circular mils wire and the other circuit is of 211,600 circular mils cable. In contrast with these figures the line of the Standard Electric Company between Electra and Mission San José, a distance of ninety-nine miles, is made up of only three conductors, each being an aluminum cable of 471,034 circular mils section. Inductance increases with the frequency of the current in a conductor, and in each of the three systems just considered the frequency is sixty cycles per second.

The use of one circuit of larger wire instead of two circuits of smaller wire has the obvious advantage of greater mechanical strength in each conductor, saves the cost of one pole line and of the erection of the second circuit. With voltages above 40,000 to 50,000 on long transmission lines there is a large loss of energy by leakage directly through the air from wire to wire. To keep this loss within desirable limits it may be necessary to give each wire of a circuit a greater distance from the others of the same circuit than can readily be had if all the wires of each circuit are mounted on one line of poles. If there is only one three-wire circuit to be provided for, three lines of poles or two lines with a long crosspiece between them may be set with any desired distance between the lines so that the leakage through the air with one wire on each pole will be reduced to a small quantity. On a line built in this way it would be practically impossible for an arc to start between the wires by any of the usual means.

Distances from pole to pole in the same line vary somewhat with the number, size, and material of the conductors to be carried. On ordinary construction in a straight line poles are often spaced from 100 to 110 feet apart—that is, about fifty poles per mile. On curves and near corners the spacing of poles should be shorter. Poles for the 80.3 miles, mentioned in New Hampshire, are regularly located 100 feet apart. Of the two pole lines between Niagara Falls and Buffalo, the older was designed to carry twelve copper cables of 350,000 circular mils each, and its poles were spaced only 70 feet apart. The newer line is designed to carry six aluminum cables of 500,000 circular mils each and its poles are 140 feet apart. Poles in each of the lines between Cañon Ferry and Butte are regularly spaced 110 feet apart and each pole carries three copper cables of 106,500 circular mils.

FIG. 81.—Chambly-Montreal Line Crossing the Chambly Canal.

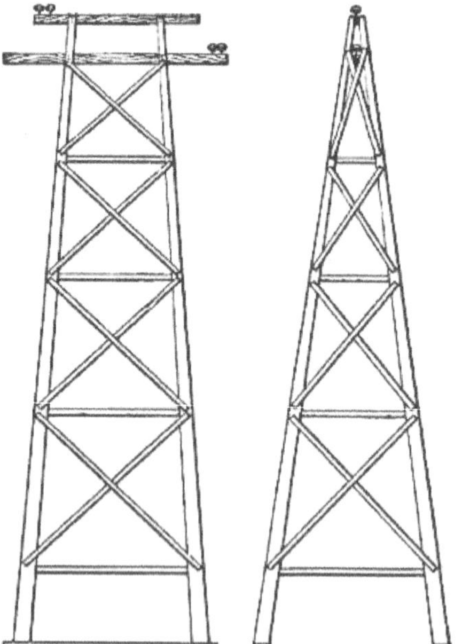

FIG. 82.—Special Wooden Structures on Line Between Spier Falls and Schenectady.

The two 142-mile lines between Colgate and Oakland are each made up of poles 132 feet apart, and one line of poles carries the three copper conductors and the other line of poles the aluminum conductors already named. As aluminum wire has only one-half the weight of copper wire of equal conductivity, the length of span between poles carrying aluminum wire may be greater than that where copper is used. Only a part of the strain on poles is due to the weight of wires carried, however. Where a body of water must be crossed, a very long span, with special supports for the wires at each side, may be necessary. A case of this sort was met where the Colgate and Oakland line crosses the Carquinez Straits at a point where the waterway is 3,200 feet wide. It was necessary to have the lowest part of the cables across these straits at least 200 feet above the surface of the water so that vessels with the tallest masts could pass underneath. To secure the necessary elevation for the cables a steel tower was built on each bank of the straits at such a point that the distance between the points for cable support on the two towers is 4,427 feet apart. As the banks rise rapidly from the water level, one steel tower was given a height of only 65 feet, while the height of the other was made 225 feet. Between these two towers four steel cables were suspended, each cable being made up of nineteen strands of galvanized steel wire, having an outside diameter of seven-eighths inch and weighing 7,080 pounds for the span. The breaking strain of each cable is 98,000 pounds, and it has the electrical conductivity of a No. 2 copper wire. The cables are simply supported on the towers by steel rollers, and the pull of each cable, amounting to twelve tons, is taken by an anchorage some distance behind each tower, where the cable terminates. Each anchorage consists of a large block of cement deeply embedded in the ground, and with anchor bolts running through it. Each cable is secured to its anchorage through a series of strain insulators, and the regular line cables of copper and aluminum are connected with the steel cables just outside of the shelter built over the strain insulators of each anchor. Steel cables were used for the long span across the straits because of the great tensile strength that could be had in that metal. This span is, no doubt, the longest and highest that has ever been erected for electrical transmission at high voltage.

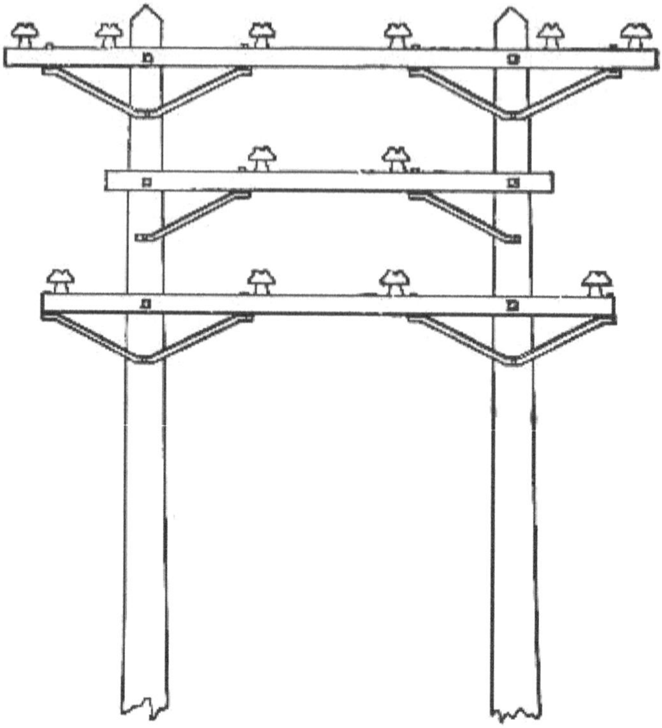

FIG. 83.—Special Structure on Line Between Spier Falls and Schenectady.

It has been suggested in one instance that steel towers ninety feet high and 1,000 feet apart be substituted for pole lines and the wires strung from tower to tower. Such construction would increase the difficulty of insulation and would cost more at the start than a line of wooden poles. The question is whether a lower maintenance and depreciation rate for the steel towers would offset their disadvantages compared with poles. Pole lines should be staked out with a transit, and the same instrument can be used to give a perpendicular position to each pole and bring it into line. Wooden poles are used in most cases of high-voltage transmission lines. Iron poles would make it unsafe to work on any circuit carried by them when it was transmitting current at high voltage. With iron poles a defective insulator might lead to the destruction of the conductors at that point through continuous arcing on to the iron.

FIG. 84.—Crossing of Delaware and Hudson Railway Tracks by 30,000-volt Line at Saratoga, N. Y.

The kinds of wood used for poles vary in different sections of the country. In New England, chestnut poles are a favorite and were used on the 80.3 miles of transmission line mentioned in New Hampshire. Cedar poles are used to some extent in nearly all parts of the country, including Canada. Spruce and pine poles are employed to some extent, especially in lengths of more than fifty feet. In the Rocky Mountain region and in California round cedar poles from the forests of Oregon, Washington, and Idaho are much used. Sawed redwood poles from the trunks of large trees were erected on the 147-mile line between Electra power-house and San Francisco. For the Colgate and Oakland line Oregon cedar poles were selected, and the transmission between Cañon Ferry and Butte was carried out with cedar poles from Idaho. For transmission circuits it is far more important at most points to have poles very strong rather than very long. Where wires or obstructions must be crossed by the high-voltage circuits the poles should be long enough to carry these circuits well above everything else. In the open country, where no obstructions are to be avoided, it does not pay to use poles with a length greater than thirty-five feet.

FIG. 85.—Pole Line from Spier Falls over Mount McGregor.

Short poles offer less surface to the wind, the length of the lever through which wind pressure acts to break the pole at the ground decreases with the length of pole, and the shorter the poles the smaller is the strain on struts and guy wires. If poles are only thirty or thirty-five feet long, they may be large in diameter without excessive cost. As a rule, no pole should be used with a top less than seven inches in diameter, and a pole with this top should not be required to carry more than three wires. A pole with seven- or eight-inch top and thirty feet long should measure not less than twelve inches in diameter at the butt. For longer poles the diameters at the butt should increase up to at least eighteen inches for a round pole sixty feet long.

In the New Hampshire transmission above named the standard length of poles is thirty-five feet. On the line between Cañon Ferry and Butte the poles range from thirty-five to ninety feet in length. The round cedar poles used in the Colgate and Oakland line range from twenty-five to sixty feet in length, from eight to twelve inches diameter at the top, and from twelve to eighteen inches diameter at the butt. On the line between Electra and San Francisco the square-sawed redwood poles are reported to have the following dimensions, in a paper read at the annual convention of Edison Illuminating Companies in 1902.

Height, Feet.	Top, Inches.	Butt, Inches.	Depth in Ground.
35	7 × 7	12 × 12	5.5
40	8 × 8	13 ½ × 13 ½	6
45	9 × 9	15 × 15	6.5
50	10 × 10	16 × 16	7
60	11 × 11	17 × 17	8

The relative dimensions of these poles are of interest because, being sawed from the trunks of large trees, they could have any desired measurements at the tops and butts. These poles, over the greater part of the line, carried the three aluminum cables of 471,034 circular mils each, previously mentioned. Depth to which poles are set in the ground ranges from about five feet for twenty-five- or thirty-foot poles to eight feet for poles sixty feet long. In locations where the soil is very soft or where poles must resist heavy strains the stability of each pole may be much increased by digging the hole two feet or more larger in diameter than the butt of the pole, and then filling in cement concrete—one part, by measure, of Portland cement, three of sand and five of broken stone—all around the butt of the pole after it is in the hole. The butts of poles up to a point one foot or more above the ground line are frequently treated with hot tar, pitch, asphalt, or carbolineum before the poles are erected, and in Salt Lake City salt is said to be used around pole butts after they are in the hole.

In some cases the poles of transmission lines are painted over their entire length. Pole tops should always be pointed or wedge-shaped, to shed water, and paint or tar should be applied to these tops. In some cases poles are filled with crude petroleum or other preservative compound in iron retorts, where moisture is withdrawn from the pole by exhausting the air, and then, after treatment with dry steam, the poles have the compound forced into them by hydraulic pressure.

In favorable soils cedar poles may remain fairly sound for twenty years, chestnut poles more than one-half of that time, and spruce and pine about five years. Poles up to forty feet in length may be readily set with pike poles, but when they are much longer than this a derrick will save time and labor. A derrick should have a little more than one-half the length of the poles to be set.

FIG. 86.—Chambly-Montreal Line Crossing the Richelieu River.

Poles should be guyed or braced at all points where there are material changes in the direction of the line, and on long straight stretches about every fifth pole should be guyed or braced in both directions to prevent the poles setting back when the line wire is cut or broken at any point. Where there is room for wooden struts, as on a private right of way, they should be used instead of guys because of their more substantial character and the higher security of insulation thus obtained. Ordinary strain insulators cannot be relied on with lines that operate at very high voltages, and where guys must be used a timber four by six inches and ten to twenty feet long may have the guy twisted about each end of it and be used as a strain insulator. A guy or strut should come well up under the lower cross-arm on a pole to avoid breaking of the pole at the point of attachment.

Where poles have heavy circuits and several cross-arms each it is sometimes desirable to attach a guy or strut beneath the lowest arm and also a guy close to the pole top. Galvanized iron or steel wire is the material best suited for guys, and the cable form has greater strength and is more flexible than solid wire.

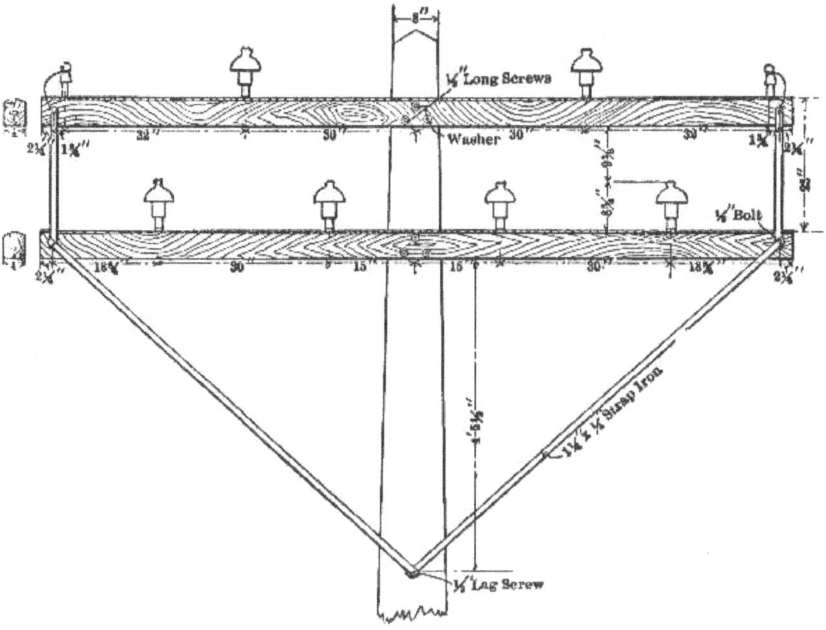

FIG. 87.—Cross-arms and Insulators on the Line Between the Chambly Plant and Montreal.

On the transmission line between Electra and San Francisco, which is intended to operate at 60,000 volts, the use of guys has been mostly avoided and struts employed instead. Where a guy had to be used, a strain insulator of wood six by six inches and twenty feet long was inserted in it.

The number and spacing of cross-arms on the poles of transmission lines are regulated by the number of circuits that each pole must carry and by the desired distance apart of the wires. Formerly it was common to carry two or more circuits on a single line of poles, but now a frequent practice is to give each pole line only one circuit and each pole only one cross-arm, except that a small cross-arm for a telephone circuit is placed some feet below the power wires. With only one transmission circuit per pole line, one wire is usually placed at the top of the pole and the other two wires at opposite ends of the single cross-arm. The older pole line for the transmission between Niagara Falls and Buffalo carried two cross-arms per pole for the power wires, these cross-arms being two feet apart. Each cross-arm was of yellow pine, twelve feet long, four by six inches in section, and intended to carry four three-wire circuits, but only two circuits have been erected on these two cross-arms. On the later pole line for this same transmission each pole carries two cross-arms, the upper intended for four and the lower cross-arm for two wires, so that one three-wire circuit may be strung on each side of the poles, two wires on the upper and one on the lower arm in the form of an equilateral triangle. The pole lines between Cañon Ferry and Butte, Colgate and Oakland, and Electra and San Francisco all have only one cross-arm for power wires per pole, and the third wire of the circuit in each case is mounted at the top of the pole so that the three conductors are at the corners of an equilateral triangle.

This relative position of the conductors makes it easy to transpose them as often as desired. On the line from Cañon Ferry to Butte the cross-arms are each eight feet long with two holes for pins seventy-eight inches apart, and are attached to the pole five feet ten and one-half inches from the top. Gains for cross-arms should be cut from one to two inches deep in poles before they are raised, and one hole for three-quarters or seven-eighths-inch bolt should be bored through the centre of the cross-arm and of the pole at the gain. Each cross-arm should be attached to the pole by a single bolt passing entirely through the pole and cross-arm with a washer about three inches in diameter next to the cross-arm. One large through bolt weakens the pole and arm less than two smaller bolts or lag-screws, and the arm can be more easily replaced if there is only one bolt to remove. Alternate poles in a line should have their cross-arms bolted on opposite sides, and at corners double arms should be used.

Yellow pine is a favorite wood for cross-arms, though other varieties are also used. The large, long pins necessary on high voltage lines tend to increase the sectional area of cross-arms, and a section less than five and one-half by four and one-half inches is seldom desirable. On the line between Electra and San Francisco, which carries the three aluminum cables of 471,034 circular mils each, the cross-arms of Oregon pine have a section of six by six inches each. Standard dimensions of some smaller cross-arms are four and three-quarters by three and three-quarters inches, but it may be doubted whether these arms are strong enough for long transmission work. Cross-arms should be surfaced all over and crowned one-quarter to one-half inch on top so as to shed water. After being kiln dried, cross-arms should be boiled in asphaltum or linseed oil to preserve the wood and give it higher insulating properties. Cross-arms longer than five feet should be secured by braces starting at the pole some distance below each arm and extending to points on the arm about half-way between the pole and each end of the arm.

FIG. 88.—Tail Race and Pole Line at Chambly, Quebec Power-station.

Each brace may be of flat bar iron about one and one-half by one-quarter inch in section, or the brace for both ends of an arm may be made of a single piece of angle-iron bent into the proper shape. For high-voltage lines it is undesirable to employ iron braces of any sort, since these braces form a path of low resistance that comes much too close to the pins on which the insulators and wires are mounted. Braces formed of hard wood are much better as to insulation, and such braces of maple are in use on the line between Butte and Cañon Ferry where the voltage is 50,000. Each brace on that line is thirty-six inches long and three inches wide, with one end bolted to the centre of its pole and the other end to the cross-arm twenty-three inches from the pole centre.

The line from Electra has hard-wood braces secured with wood pins.

Wood is the most common material for pins on which to mount the insulators of high-voltage transmission circuits. Iron has been used for pins to some extent, and its use is on the increase. Oak and locust pins are generally used, the latter being stronger and more lasting. In California, pins of eucalyptus wood are much used and are said to be stronger than locust. All wooden pins should be boiled several hours in linseed oil after being well dried. This increases the insulating and lasting properties of the pins.

High-voltage lines require long pins to hold the lower edges of insulators well above the cross-arms, and these pins must be much stronger than those used on ordinary lines, because of the increased leverage of each wire.

A pin twelve inches long over all and having a diameter of one and one-half inches in the part that enters the cross-arm has been much used for transmission circuits, but is much too short and weak for high voltages. On the 50,000-volt line between Cañon Ferry and Butte the pins are seasoned oak boiled in paraffin. Each of these pins is seventeen and one-half inches long, two and one-half inches in diameter for a length of four and one-half inches in the middle part, two inches in diameter for a length of five and one-half inches that fits into the cross-arm or pole top, and one and one-half inches in diameter at the top of the thread inside of the insulator. These pins hold the outside edges of the insulators nine inches above the tops of cross-arms. Each of these pins is held in its socket by a three-eighths-inch bolt that passes entirely through the pin and the cross-arm or pole top.

On the line between Electra and San Francisco the pins are each sixteen and seven-eighths inches long, two and three-quarters inches in diameter at the largest central part, and two and one-quarter inches in diameter in the lower part, five inches long, that fits into the cross-arm or pole top. One of these pins broke at the shoulder with a pull of 2,200 pounds at the threaded part. Carriage bolts one-half inch in diameter pass through the cross-arm and pin two inches from the top of the arm, and one bolt three inches from the pin on each side. Without these bolts the arms split on test with a pull of 1,200 pounds on the pin, but with the bolts the pin broke as above.

CHAPTER XIX.
ENTRIES FOR ELECTRIC TRANSMISSION LINES.

The entrance of transmission lines into generating plants and sub-stations presents special problems in construction and insulation. One of these problems has to do with the mechanical security of each conductor at the point where it passes through the side or roof of the station. Conductors are sometimes attached to the station so that the strain of the line is borne by the side wall where they enter and tends to pull it out of line.

This practice has but little to commend it, aside from convenience, for unless the conductors are rather small, or the wall of the station is unusually heavy, the pull of the former is apt to bulge the latter in the course of time. For any heavy line the end strain is ultimately most suitably taken by an anchor securely fixed. As special insulators must be used where a conductor is secured directly to such an anchor, it is usually more convenient to set one or more heavy poles with double cross-arms at the end of a line, and then to make these poles secure by large struts, or by guys attached to anchors. Extra heavy cross-arms on these end poles should be provided with iron pins for the line insulators; two or more of the insulators mounted in this way within a few feet of each other, for each wire, will stand up against the end strain on almost any line.

Insulators that are to take the end strain of a line in this way should allow attachment of the wire at the side, so that the force exerted by each conductor tends to press the insulator against the side of its pin, rather than to pull off the top of the insulator. The end strain of the line having been taken on poles close to the station, the conductors may be attached to insulators on the wall, the latter thus being subjected to very little mechanical strain.

Overhead lines usually enter a station through one of its side walls, but an entry may be made in the roof. It is desirable to have a side entry on the gable end of a building rather than on a side below the eaves where there will be much dripping of water. If an entry must be made below the eaves, a shelter should be provided above the entry, and the roof of this shelter should have a gutter that will carry water away from the wires.

Entrance of each conductor into a station must be effected in such a way that ample insulation of the circuit will be maintained, and in some cases so that rain, snow, and wind will be excluded. The line voltage and the climate where the station is located thus have an important bearing on the form of entry that is suitable in any particular case.

The simplest form of entry for a high-voltage line is a clear opening, usually circular in form, through the wall of the station for each wire.

Insulators for each wire should be provided both inside and outside of the wall to hold the wire at the centre of this opening. Such insulators are usually most conveniently supported by fixtures attached to both sides of the wall, and insulators on the outside should of course be kept in an upright position, unless completely protected from rain and snow.

The diameter of the openings through the wall should be great enough to prevent any visible discharge of current between the wire and wall under the worst conditions of snow, rain, fog, or dust. Such an opening must, therefore, increase in diameter with the voltage of the line. The larger these openings for the line wires, the greater is the opportunity for rain, snow, dust, and cold air to enter the station through them.

Openings may be so protected as to keep out snow and rain by means of shelters on the outside of the wall on which they are placed, but such shelters cannot keep out the cold air. If the openings for the entrance of wires are located in the wall of a room that contains air-blast transformers, the area of openings for circuits of very high voltage may be no greater than is necessary to allow the escape of heated air from the transformers.

The milder the climate, other factors being the same, the higher the voltage of circuits which may enter a station through openings that are free for the movement of air. With circuits of only moderate voltage, say less than 15,000, it is quite practicable to admit wires to a station through perfectly free openings, in the coldest parts of the United States. With voltages of 20,000 to 60,000 it is often necessary, in the colder parts of the country, to close the opening in the wall through which each wire enters with a disc of insulating material.

In order to keep the current leakage over these discs within proper limits, the diameters of the discs must increase with the voltage of the circuit. This increase of disc diameter obviously lengthens the path of leakage current over the disc surface. Where the openings in a wall for the entrance of high-voltage circuits are closed by insulating discs about the wires, these discs may make actual contact with bare wires, or the wire at each entry may have some special insulation.

In the side wall of the sub-station at Manchester, N. H., the entrance of transmission lines from four water-power plants is provided for by circular openings in slate slabs that are built into the brickwork. The transmission circuits from three of the water-power plants operate at 10,000 to 12,000 volts, and the circuit from the fourth plant at about 6,000 volts. Circular openings in the slate slabs are each five inches in diameter, and they are spaced twelve to fifteen inches between centres. A single wire enters through each of these openings and is held at the centre by insulators both inside and outside of the wall. Each wire is bare where it passes through the slate slab, and the circular openings are not closed in any way. The largest wires passing through these five-inch circular openings in the slate slabs are of solid copper, No. 0, of 0.325-inch diameter each.

Before passing through the opening in the slate slabs the wires of these transmission circuits are tied to regular line insulators supported by cross-arms secured to the outside of the brick wall by iron brackets. The point of attachment of each wire to its insulator is about nine inches below the centre of the circular hole by which it enters the sub-station.

This Manchester sub-station is equipped with air-blast transformers from which the hot air is discharged into the same room that the transmission lines enter. Along one side of the sub-station there are twenty-seven of these five-inch circular openings in the slate slabs for entrance of the high-voltage lines, and on another side of the sub-station there are a greater number of smaller openings for the distribution circuits. Were it not for the air-blast transformers, all of these openings would probably admit more air than would be desirable in a climate as cold as that at Manchester.

Another example of openings in the walls of a station for the entrance of transmission circuits, where there is free movement of the air between the inside and outside of the building, is that of the 33,000-volt line between Santa Ana River and Los Angeles, Cal. In this case a sewer pipe twelve inches in diameter is built into the wall of the station for each wire of the line, so that there is a free opening of this size from inside to outside.

Each wire of the 33,000-volt circuit enters the station through the centre of one of these twelve-inch pipes, and is thus surrounded by six inches of air on every side. As the temperature near Los Angeles seldom or never goes down to zero, these large openings do not admit enough air to be objectionable. Besides this mild climate, air-blast transformers add to the favorable features in the stations having the twelve-inch openings.

In another case, however, where the openings for the entrance of wires of very high voltage allow free movement of air between the inside and outside of the station, the climate is cold and the winter temperatures go down to 30° or more below zero. This condition exists on the 25,000-volt line between Apple River Falls and St. Paul, where six No. 2 wires enter the generating station through plain circular openings in the brick side wall of a small extension where the lightning arresters are located. Air-blast transformers are located in the end of the station next to this lightning-arrester house, but it is not certain that the hot air from them escapes through the openings for the wires.

In another case where the climate is about as cold as that just named, a gallery is built along one side of the exterior of the station at some distance above the ground, and two openings are provided for each wire of the high-tension line. One of these two openings is in the horizontal floor of the gallery and allows the entrance of the wire from the outside, and the other opening is in the side wall of the station against which the gallery is built. The two openings for each wire being thus at right angles to each other, and the opening to the outside air being protected from the wind by its horizontal position, no more than a permissible amount of cold air, it is said, finds its way into the station.

In some cases with lines of moderate voltage, say 10,000 to 15,000, and in probably the majority of cases with lines of 25,000 volts or more, the entry for the high-tension wires is entirely closed. An example of this practice may be seen at the various sub-stations of the New Hampshire Traction Company, which are located along their 12,000-volt line between Portsmouth and Pelham, in that State.

For the entry of each wire on these lines a sixteen-inch square opening is made in the brick wall of the sub-station. On the outside of this wall a box is built about a group of three or more of these openings located side by side. The top or roof of this box is formed by a slab of bluestone three inches thick, which is set into the wall and extends twenty-six inches from the face of the wall, with a slight slope from the horizontal.

The ends, the bottom, and the outer side of this box are formed by slabs of slate one inch thick, so that the enclosed space has an area in vertical cross section at right angles to this building 15.5 inches high and twenty-two inches wide.

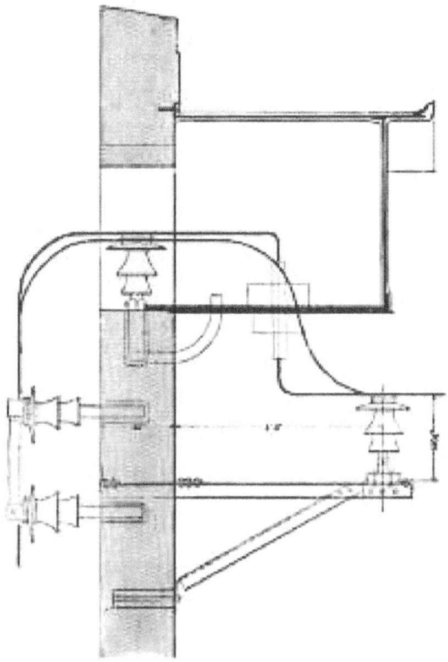

FIG. 89.—Cable Entering Building.

In the bottom of this box there is a circular opening for each wire, and into this opening fits a heavy glass or porcelain bushing through which the wire passes. After reaching the inside of the box the wire turns at right angles and passes through the sixteen-inch square opening into the sub-station. Beneath the box a special insulator is secured by an iron bracket to the outside of the brick wall for each line wire, and this insulator takes the strain of the wire before it is carried up through the bushing in the bottom of the box. This form of entry is permissible where the desire is to exclude cold air from the station, and where the voltage is not high enough to cause serious leakage over the surface of the bushing and the slate forming the bottom of the box. In all of the cases above mentioned the wires used to enter the stations were the regular line conductors and were bare.

Another type of entry in sub-stations is that employed on the extensive transmission system between Spier Falls, Schenectady, and Albany, N. Y. The maximum voltage on this system is 30,000, and the lines usually enter each sub-station through the brick wall at one of its gable ends. Outside of and about the entry of each circuit or group of circuits a wooden shelter is built on the brick wall of the sub-station. Each shelter has a slanting roof that starts from the brick wall at some distance above the openings for the entrance of the line, and terminates in a gutter. The front of each shelter is carried down three feet below the centre of the openings in the brick wall, and the ends go still lower. The front of each shelter is four feet in height, is four feet from the face of the brick wall, and has a circular opening of 10-inch diameter for each wire of the transmission line.

In line with each circular opening in the wooden shield there is an opening of 15-inch diameter in the brick wall of the sub-station, and into this opening in the brickwork fits a ring of wood with 15-inch outside and 11-inch inside diameter. To this wooden ring a 15-inch disc of hard fibre ⅛-inch thick is secured, and a porcelain tube 24 inches long and of 2-inch inside diameter passes through a hole in the centre of this disc. Within the wooden shield and in line with each circular opening in it and with the corresponding porcelain tube through the fibre disc a line insulator is secured. Within the sub-station and in line with each tube there is also an insulator, and the two insulators near opposite ends of each tube hold the line wire that passes through it in position.

Each wire of the transmission lines, of which the largest is No. 000 solid of 0.410-inch diameter, terminates at one of the insulators within the wooden shield, and is there connected to a special insulated wire that passes through one of the porcelain tubes into the sub-station. A copper trolley sleeve 12 inches long is used to make the soldered connection between the bare line wire and the insulated conductor that passes through the porcelain tube. Each of these entry cables, whatever its size, is insulated first with a layer of rubber 9/32-inch thick, then with varnished cambric wound on to a thickness of 9/32-inch, and lastly with two layers of weather-proof braid outside of the cambric. This form of closed entry for the transmission lines obviously excludes snow, rain, cold air, and dust from the station. Whether the fibre discs and wooden rings, together with the insulation on the entry cables, are as desirable as a glass disc at the entry is another question.

Another instance where the entry for a high-tension line is closed with the aid of combustible material is that of the 25,000-volt transmission between the water-power plant at Chambly, on the Richelieu River, and the sub-station in Montreal. The four three-phase circuits of this line are made up of No. 00 wires of 0.365-inch diameter each, which enter the power-station at Chambly and the terminal-house in Montreal bare, as they are outside.

At each end of the line the wires are secured to insulators on a horizontal arm with their centres twenty-two inches outside of an end wall of the station or terminal building. The insulators are mounted with their centres thirty inches apart, and a few inches above the tops of these insulators a corresponding row of wooden bushings pass through the wall with an outward slant.

At the Chambly end of the line each of these bushings is of oak, boiled in stearin, four inches in diameter and twelve inches long. At the Montreal end the wall bushings are of boxwood, and each is four inches square and twelve inches long. Each of the wooden bushings carries a glass tube, and is itself held in position by the concrete of the wall in which it is located. Entrance to the station by each of the bare No. 00 wires is gained through one of these glass tubes, and cold air is excluded.

Quite a different type of closed entry for the wires of a transmission line is in use on that between Shawinigan Falls and Montreal, which operates at 50,000 volts. For the entry of each of the three aluminum cables that make up this line, each cable being composed of seven No. 6 B. & S. gauge wires, a tile pipe of twenty-four-inches diameter was set into the station wall. The end of each tile pipe is closed by a glass plate, with a small hole at its centre, through which the cable passes.

As the cable is thus held twelve inches from the terra cotta pipe all the way around, any leakage of current must pass over this length of glass surface at each cable or through the air.

A heavy coating of frost sometimes collects on these plates, and this increases the amount of current leakage over them. Surface leakage in a case of this sort, of course, varies with the size of the glass plate, and if a tile pipe is used the limit of size is soon reached.

There seems to be no good reason, however, why a glass plate of any desired dimensions should not be set directly into the brick wall of a station for each line wire and the tile pipes entirely omitted. This plan is followed on the system of the Utah Light & Power Company, which extends to Salt Lake City, Ogden, Provo, and a number of other points in that State.

On the 40,000-volt line of that system an entry for each wire is provided by setting two plates of glass into the brick wall, one plate being flush with the inner surface and the other with the outer surface of the wall.

In the centre of each plate there is a hole of about 2.5-inch diameter, into which a glass or porcelain tube fits. The line wire enters the station through this tube, and it does not appear that any shelter for the glass plates is located outside of the building. An entry of this type for the 40,000-volt line with glass plates in a brick wall at a gable end of the Murphy mill is said to have given satisfactory results during four years, though that wall faces the southwest, from which direction most of the storms come. At this entry each glass plate is not more than eighteen inches in diameter, and the wires are about four feet apart. On a 16,000-volt line of the same company, a glass plate twelve inches square with a three-quarter-inch hole at its centre, and the bare wire passing through without a tube, has given results that were entirely satisfactory.

Two quite different types of entry to stations are used on the 50,000-volt line between Cañon Ferry and Butte, Mont. One type, employed at the side wall of a corrugated iron building, consists of a thick bushing of paraffined wood carrying a glass tube two inches in diameter, four feet long, with a side wall of five-eighths to three-quarter-inch, through which the line conductor passes.

On the roof of the power-station at Cañon Ferry a vertical entry is made with the 50,000-volt circuit. For this purpose each line wire is brought to a dead end on three insulators carried by a timber fixture on the roof. A vertical tap drops from each line wire and passes through the roof and into the station. This roof is of wood, covered with tin outside and lined with asbestos inside. Each tap is an insulated wire, and elaborate methods are adopted in the way of further insulation, and to prevent water from following the wire down through the roof.

Over the point of entrance sits a large block of paraffined wood with a central hole, and down through this hole passes a long cylinder of paper that extends some distance above the block. Into the top end of this cylinder fits a wood bushing, and a length of the tap wire that has been served with a thick layer of rubber is tightly enclosed by this bushing. The rubber-covered portion of the tap wire also extends above the bushing, and has taped to it a paper cone that comes down over the top of the paper cylinder to keep out the water. On the outside of this paper cylinder, at a lower point, a still larger paper cone is attached to prevent water from following the cylinder down through the wooden block. At the lower end of the paper cylinder, within the station, there is another bushing of wood, and between this and the wooden bushing at the top of the cylinder and inside of the paper cylinder there is a long glass tube. Down through this tube and into the station the insulated tap wire passes.

From the experience thus far gained with high-voltage lines, it seems that their entrance into stations should always be at a side wall, unless there is some imperative reason for coming down through the roof. If climatic conditions permit, no form of entry can be more reliable than a plain, ample opening through the wall with a large air-space about each wire. If the opening must be closed, it had better be done with one or more large plates of thick glass set directly into the brickwork of the wall. Some additional insulation is obtained by placing a long glass or porcelain tube over each wire where it passes through the central hole in the glass plates. Each conductor should be bare at the entry, as it is on the line. Some of the above examples of existing practice in entries for transmission lines are taken from Vol. xxii., A. I. E. E.

CHAPTER XX.
INSULATOR PINS.

Wooden insulator pins are among the weakest elements in electric transmission systems. As line voltages have gone up it has been necessary to increase the distances between the outside petticoats of insulators and their cross-arms and to lengthen the insulators themselves in order to keep the leakage of current between the conductors within permissible limits. To reduce the leakage, the wires on most lines are located at the tops instead of in the old position at the sides of their insulators.

All this has tended to a large increase of the mechanical strains that operate to break insulator pins at the point where they enter the cross-arm, because the strain on each line wire acts with a longer leverage. Again, it is sometimes necessary that transmission lines make long spans across rivers or elsewhere, and a very unusual strain may be put on the insulator pins at these places.

As long as each electric system was confined to a single city or town a broken insulator pin could be quickly replaced, and any material interruption of service from such a cause was improbable. Where the light and power supply of a city, however, depends on a long transmission line, as is now the case in many instances, and where the line voltage is so great that contact between a wire and a cross-arm will result in the speedy destruction of the latter by burning, a broken pin may easily lead to a serious interruption of the service.

Besides the increase of mechanical strains on insulator pins, there is the danger of destruction of wooden pins by charring, burning, and other forms of disintegration due to leakage of current over the insulators. This danger was entirely absent in the great majority of cases so long as lines were local and operated at only moderate voltages. These several factors combined are bringing about marked changes in design.

On straight portions of a transmission line the insulator pins are subject to strains of two principal kinds. One of these is due directly to the weight of the insulators and line wire, and acts vertically to crush the pins by forcing them down onto the cross-arm. The other is due to the horizontal pull of the line wire, which is often much increased by coatings of ice and by wind pressure, tending to break the pins by bending—most frequently at the point where they enter the cross-arm. A strain of minor importance on pins is that encountered where a short pole has been set between two higher ones, and the line at the short pole tends to lift each insulator from its pin, and each pin from the cross-arm.

Where the line changes its direction, as on curves and at corners, the side strain on pins is greatly increased, and such places give by far the largest amount of trouble through the breaking of pins. The latter seldom fail by crushing through the weight of the lines they support, because the size of pin necessary to withstand the bending strain has a large factor of safety as to crushing strength. Insulators are sometimes lifted from wooden pins, and the threads of these pins stripped where a short pole is used, as already noted; but failure of this kind is not common.

Iron pins are either screwed or cemented into their insulators, but the cemented joint is much more desirable, because where a screw joint is made the unequal expansion of the iron and the glass or porcelain is apt to result in breakage of the insulator. Where cement is used, both the pins and insulators should be threaded or provided with shoulders of some sort, so that, although the shoulders of threads do not come into contact with each other, they will, nevertheless, help to secure a better hold. Pure Portland cement, mixed with water to a thick liquid, has been used with success, the insulator being placed upside down and the pin held in a central position in the hole of the insulator while the cement is poured in. Another cement that has been used for the same purpose is a mixture of litharge and glycerin. Melted sulphur is also available.

The same forces that tend to lift an insulator from its pin tend also to pull the pin from its socket in the cross-arm or pole top. With wooden pins the time-honored custom has been to drive a nail into the side of the cross-arm so that it enters the shank of the pin in its socket. This plan is good enough so far as immediate mechanical strength is concerned, but is not desirable, because it is hard to remove a nail when a pin is to be removed, and also because the rust of the nail rots the wood. A better plan is to have a small hole entirely through each cross-arm and insulator pin at right angles to the shank of that pin in its socket, and then to drive a small wooden pin entirely through from side to side.

Some of the important factors affecting the strains on insulator pins vary much on different transmission lines, as may be seen from the following table of lines on which wooden pins are used. On the older line between Niagara Falls and Buffalo, the regular length of span is 70 feet, and each copper conductor of 350,000 circular mils is attached to its insulator 7.5 inches above the cross-arm. On the new line the length of span is 140 feet, and each aluminum conductor of 500,000 circular mils is attached to its insulator 10 inches above the cross-arm.

TABLE I.—DATA OF LINES ON WOODEN PINS.

Location of the Lines.	Circular Mils of Each Conductor.	Feet Length of Span Between Poles.	Inches from Wire to Shank of Pin.
Colgate to Oakland	[B]133,100	...	13
Electra to San Francisco	[A]471,034	130	15
Cañon Ferry to Butte	[B]105,600	110	13 ½
Shawinigan Falls to Montreal	[A]183,750	100	16 ¼
Niagara Falls to Buffalo	[B]350,000	70	7 ½
Niagara Falls to Buffalo	[A]500,000	140	10
Chambly to Montreal	[B]133,100	90	8 ½
Colgate to Oakland	[A]211,600	...	13

[A] Aluminum conductors.

[B] Copper conductors.

TABLE II.—DIMENSIONS OF WOODEN PINS IN INCHES.

Location of Lines.	Length of Stem.	Length of Shank.	Diameter of Shank.	Diameter of Shoulder.	Diameter of Threaded End.	Length of Threaded Part.
Colgate to Oakland	10 ⅜	5 ⅜	2 ⅛	2 ½	1 ⅜	2
Electra to San Francisco	12	4 ⅞	2 ¼	2 ¾	1 ⅜	2
Cañon Ferry to Butte	12 ½	5 ⅛	2	2 ½	1 ⅛	3
Shawinigan Falls to Montreal	13 ½	5	2 ¾	3	1	..
Niagara Falls to Buffalo[A]	5 ¼	6	2	2 ¾	⅞	1 ½
Niagara Falls to Buffalo[B]	7 ¾	6	2 ¼	2 ¾	1 ½	2 ½
Chambly to Montreal[C]	7	5	1 ½	1 ⅞	..	..

Cañon Ferry to Butte[D]	12 ⅜	7 ⅞	2 ⅛	2 ½	1 ⅛	3

[A] Pins on old line.

[B] Pins on new line.

[C] Approximate dimensions.

[D] Pole top pins.

To compensate for the greater strains introduced by doubling the length of span and using pins of longer stem, the diameter of the shank of the new pins was increased to two inches. One line between Colgate and Oakland is of copper, and the other is of aluminum conductors, but the same pins appear to be used for each. On the line between Cañon Ferry and Butte, Mont., the pin used in pole tops has a shank 2¾ inches longer and ⅛-inch greater in diameter than the pin used in cross-arms. The weakest pin included in the table seems to be that in use on the line between Chambly and Montreal, which is of hickory wood, about 1½ inches in diameter at the shank, and carries its No. 00 copper wire 8½ inches above the cross-arm.

The following dimensions for standard wooden insulator pins to be used on all transmission lines are proposed in vol. xxi., page 235, of the Transactions of the American Institute of Electrical Engineers. These pins are designed to resist a uniform pull at the smaller end and at right angles to the axis in each case. The length of each pin, in inches between the shoulder and the threaded end, is represented by L, and the diameter of each pin at its shank by D.

L.	D.
1	0.87
2	1.10
3	1.26
4	1.39
5	1.50
6	1.59
7	1.67
8	1.75
9	1.82
10	1.88
11	1.95

13	2.06
15	2.17
17	2.25
19	2.34
21	2.42

The two strongest pins in Table II. appear to be those in use on the line between Shawinigan Falls and Montreal and on the line from Niagara Falls to Buffalo. The former have a diameter of 2¾ inches at the shank, and the wire is carried 16¼ inches above the shoulder of the pin. On the new Niagara line the shank diameter of each pin is only 2¼ inches, but the line wire is only 10 inches above the shoulder. It was found by tests that a strain of 2,100 pounds at the top of the insulator and at right angles to the axis of this Niagara pin was necessary to break it at the shank. This strain is about six times as great as the calculated maximum strain that will occur in service on the line.

Some of the pins here noted are much stronger than those proposed in the above specifications for standard pins. The pins on the old Niagara line have a shank diameter of 2 inches, with a stem only 5¼ inches long, while the proposed pin of 2 inches diameter at the shank has a stem 11 inches long. On the Colgate and Oakland line a shank diameter of 2⅛ inches goes with a length of 10⅜ inches in the stem, but the proposed pin with this size of shank has a stem 13 inches long. For a shank of 2¼ inches diameter the proposed pin has a stem 15 inches long, but the pins with this diameter of shank on the Electra line are only 12 inches long in the stem.

The 2¼-inch diameter of shank in the pins on the new Niagara line goes with a length of only 7¾ inches in the stem. The new Niagara pin is thus almost exactly twice as strong as the proposed pin, since the strength of a pin where the shank joins the stem varies inversely as the length of the stem, all other factors being the same.

Pins on the Shawinigan Falls line have a shank 2¾ inches in diameter, with a length of 13½ inches in the stem; but the largest of the proposed pins, that with a stem 19 inches long, has a diameter of only 2½ inches in the shank.

It is hardly too much to say in the interest of good engineering that the wooden pin of about 5 inches length of stem and 1½ inches diameter of shank, as well as all longer pins of no greater strength, should be discarded for long transmission lines of high voltage. These pins have done good service on telegraph and telephone lines, and on local lighting circuits of No. 6 B. & S. gauge wire or smaller, and they may well be left for such work.

To meet the conditions of transmission work a change in both the shape and size of pins is necessary. In the first place, the shoulder on pins where the shank and stem meet, that relic of telegraph practice, should be entirely discarded. This change will save considerable lumber on pins of a given diameter at the shank, and will increase the strength of the pin by avoiding the sharp corner at the junction of the shank and stem.

Another change of design should leave an excess of strength in the stem of the pin, to provide for deterioration of the wood, and particularly for charring by current breakage. This increase of diameter and strength near the top of the pin will cost nothing in lumber, for the wood is necessarily there unless it is turned off. The shank of each pin should be proportionately shorter than in the older type, and the pin hole should be bored only part way through the cross-arm. A saving in lumber for pins and for cross-arms will thus be made, since the size of the cross-arm may be less for a given resistance to splitting.

With these changes in general design the pin is a simple cylinder in the shank, with a gentle taper from the shank to form the stem. An example of this design, which might well serve as a basis for a line of standard pins, would be a pin 2 inches in diameter and 3½ inches long in the shank, and tapering for a length of 5 inches from the shank to form the stem, with a diameter of 1½ inches at the top. The hole in a cross-arm for this pin should be 3½ inches deep, and this, in an arm 4¾ inches deep, would leave 1¼ inches of wood below the pin. From the lower end of the pin hole, a hole ¼-inch in diameter should run to the bottom of the cross-arm to drain off water. A line of longer pins designed to resist the same line pull as this short one would be strong enough for small conductors, say up to No. 1 B. & S. gauge wire.

For larger wires, long spans and sharp angles in a line, a pin 2¼ inches in diameter and 4½ inches long in the shank, tapering for 5 inches to a diameter of 1¾ inches at the top, or longer pins of equal strength, should be used.

Where the pin holes do not extend through the cross-arm there is no need of a shoulder on the pin to sustain the weight of the line wire. In the cross-arm on the new Niagara Falls line each pin hole is bored to a depth of 5 inches, leaving 1 inch of wood below the hole. On the line from Electra to San Francisco the depth of each pin hole is again 5 inches, and the depth of the cross-arm 6 inches.

The pins for use on the Electra line were kept for several hours in a vat of linseed oil at a temperature of 210° F. The pins for the Shawinigan line were boiled in stearic acid. All wooden pins should be treated chemically, but the object of this treatment should be to prevent decay rather than to give them any particular insulating value.

The lack of strength in wooden pins and their destruction in some cases by current leakage are leading to the use of iron and steel pins. Such a pin, in use on the lines of the Washington Power Company, of Spokane, Wash., is made up of a mild steel bar $17\frac{1}{2}$ inches long and $1\frac{1}{8}$ inches in diameter, cast into a shank at one end, so that the total length is 18 inches. The cast-iron shank has a diameter of $2\frac{1}{16}$ inches, with a shoulder of $2\frac{1}{2}$ inches diameter at its upper end. To prevent the pin from lifting out of its hole a small screw enters the top of the cross-arm and bears on the top end of the shank. Above the cast-iron shank the length of the steel rod is 12 inches, and starting $\frac{3}{4}$ inch down from its top a portion of the rod $\frac{3}{4}$ inch long is turned to a diameter of one inch.

It is said that this pin begins to bend with a pull of 1,000 pounds at its top, but that it will support the insulator safely even when badly bent.

Insulators may resist puncture and prevent surface arcing from wire to pin, but still allow a large though silent flow of energy over the pins and cross-arms between the conductors of a transmission circuit. The rate at which current flows from one wire of a transmission circuit to another in this way depends on the total resistance of each path over insulator surfaces and through air to the pins and cross-arm, and then over these parts.

If the pins and cross-arm are entirely of iron, the total resistance of the path through them from wire to wire is practically that of the insulator surfaces. If the pins and cross-arm are of wood which is dry, they may offer an appreciable part of the total resistance of the path through them between the wires of a circuit; but if the wood be wet, its resistance is very much reduced.

The resistance of wooden pins and cross-arm may be so small compared with that of the air and insulator surfaces that complete the path through them from wire to wire of a circuit, that the effect of these wooden parts in checking the flow of current between conductors is relatively unimportant, and yet the resistances of these pins and the cross-arm may affect their lasting qualities.

The current that flows over the pins and cross-arms from one wire to another of a high-tension circuit may be so small as not to injure these wooden parts when evenly distributed over them, and yet this same current may char or burn the wood if confined to a narrow path. Such a leakage current will naturally cease to be evenly distributed over pins and their cross-arms when certain portions of their surfaces are of much lower resistance than others, because an electric current divides and follows several possible paths in the inverse ratio of their resistances.

These narrow paths of relatively low resistance along wooden pins and cross-arms are heated and charred by the very current that they attract, so that the conductivity of the path and the heat developed therein react mutually to increase each other, and tend toward the destruction of the wood.

Among causes that tend to make some parts of pins and cross-arms better conductors than others, there may be mentioned cracks in the wood, where dirt and moisture collect, dust, with a mixture of salt deposited on the wood by the winds at certain places, and sea fogs that are often blown only against one side of the pins and arms and deposit salt.

To make matters worse, the same cause that creates a path of relatively good conductivity along wooden pins and cross-arms often materially lowers the resistance offered to the leakage of current by the insulator surfaces. Thus an increase of the rate at which energy passes from wire to wire of a circuit, and the concentration of this energy in certain parts of the wooden path, are sometimes brought about at the same time. Where the line insulators employed are so designed that the resistance of the dry wooden pins and cross-arms forms a material part of the total resistance between the wires of a circuit, a rain or heavy fog may cause a very large increase in the rate at which energy passes over these wooden parts between the conductors.

As long as only moderate voltages were carried on line conductors, the charring and burning of their pins and cross-arms was a very unusual matter; but with the application of very high pressures on long circuits the destruction of these wooden parts by the heat of leakage currents has become a serious menace to transmission systems. Even with low voltages there may be charring and burning of pins and cross-arms if the line insulators are very poor or if the conditions as to weather and flying dust are sufficiently severe.

In vol. xx. of the Transactions of the American Institute of Electrical Engineers, pages 435 to 442 and 471 to 479, an account of the charring and burning of pins on several transmission lines is given, from which some of the following examples are taken.

In one case a line that ran near a certain chemical factory was said to be much troubled by the burning of its pins, though the voltage employed was only 440, and the insulators were designed for circuits of 10,000 volts. In rainy weather, when insulators, pins, and cross-arms were washed clear of the chemical deposits, there was no pin burning. Similar trouble has been met with on sections of the 40,000-volt Provo line, in Utah, where dust, mixed with salt, is deposited on the insulators, pins, and cross-arms. On page 708 a 2,000-volt line is mentioned on which fog, dust, and rain caused much burning of pins.

When circuits are operated at voltages of 40,000 to 60,000, no very severe climatic conditions are necessary to develop serious trouble in the wooden pins by leakage currents, even where the transmission lines are supported in insulators of the largest and best types yet developed. Striking examples along this line may be seen in the transmission systems between Colgate and Oakland, Cal., and between Electra and San Francisco. Both of these systems were designed to transmit energy at 60,000 volts, but the actual pressure of operation seems to have been limited to about 40,000 volts during much of their period of service.

Insulators of a single type and size are used on both of these transmission lines, and are among the largest ever put into service on long circuits. Each of these insulators is 11 inches in diameter, and 11¼ inches high from the lower edge to the top, the line wire being carried in a central top groove. The wooden pins used on the two lines vary a little in size, so that on the Electra line each pin stands 11½ inches above its cross-arm, while on the Colgate line the corresponding distance is 12 inches. As the insulators are of the same size in each case, the length of the pin between the lower edge of each insulator and the top of the cross-arm is 4 inches on the Colgate line and 3½ inches on the Electra line.

On the latter line a porcelain sleeve, entirely separate from and making no contact with the insulator, covers each pin from the top of its cross-arm to a point above the lower edge of the insulator. On the Colgate line each insulator makes contact with its pin for a length of 2½ inches down from the top of its thread, and on the Electra line the contact of each insulator with its pin runs down 3½ inches below the top of the thread. This leaves 9 inches in the length of the pin between the insulator contact and the top of each cross-arm on the Colgate line, and a corresponding length of pin amounting to 8½ inches on the Electra line. Of this 8½ inches of pin surface, about 6 inches is covered by the porcelain insulating sleeve used on each pin of the Electra line, so that only about 2½ inches of the length of each pin on that line is exposed to the leakage of current from the insulator directly through the air. Both the sizes of pins just mentioned were made of eucalyptus wood, boiled in linseed oil.

Each one of three pins taken from a pole, between North Tomer and Cordelia, on the Colgate line, was badly charred and burned on its side that faced the damp ocean winds. This charring extended all the way down each pin from the point where the insulator made contact with it, a little under the threads, to the top of the cross-arm nine inches below. Two of these pins were located at the opposite ends of a cross-arm, and the third was fixed in the top of the pole. This cross-arm was charred or burnt, as well as the pin, but no defects could be detected in the insulators that the pins supported.

As to these three pins, the most reasonable explanation seems to be that enough current leaked over both the outside and inside surfaces of each insulator and through the air to char the pin and cross-arm. In flowing down each pin, the current was naturally concentrated on the side exposed to the damp winds of the ocean, because the deposit of moisture by these winds lowered the resistance on that side. When these winds were not blowing, and before a pin became charred on one side, its resistance was probably about the same all the way around, and the leakage current, being distributed over the pin, was not sufficient to char it. The damp wind would, of course, lower the surface resistance of each insulator, and this, with the deposit of moisture on the pins and cross-arm, many have made a very material reduction in the total resistance from wire to wire.

The insulators used on these pins each had two petticoats, an upper one, 11 inches in diameter, and a lower one, 6½ inches in diameter, the lower edge of the smaller petticoat being 7½ inches beneath the lower outside edge of the larger petticoat. As the inner surface of the larger petticoat was much nearer to a horizontal plane than the inner surface of the smaller petticoat, moisture would have been more readily retained on it, and the greater part of the surface resistance of the insulator during wet weather must therefore have been on the inside of the smaller petticoat. At its lower edge the smaller petticoat was distant radially about 1¾ inches from the pin, and the distance between the pin and the inside surface of the smaller petticoat gradually decreased to actual contact at a point 5½ inches above this lower edge.

The path of the current from the line wire to the pin in this case seems to have been first over the entire insulator surface to the lower edge of the smaller petticoat and then partly up over the inner surface of this petticoat and partly from that surface through the air. On each of these three pins the charring was quite as bad just below the thread as it was further down, so that a large part of the leakage current seems to have gone up over the interior surface of the smaller petticoat. The charred portion of these pins extended but little, if at all, into the threads near the tops or into the part of the pin fitting into the cross-arm. The preservation of the part of each pin that entered the cross-arm seems to have been due to the increase of surface and decrease of resistance of the cross-arm in comparison with the pin. Preservation of the threaded part of each pin seems to have been due to its protection from moisture and its high resistance, so that little or no current passed over it.

Another pin taken from the same line as the three just considered was badly burned at a point about 1.75 inches below the threads, but on sawing it completely across at two points below the charred spot the entire section was found to be perfectly sound and free from any sign of burning. The explanation of the condition of this pin is, perhaps, that the resistance of the burned part, owing to its additional protection and dryness, was high compared with that of the lower part of the pin, and thus developed most of the heat on the passage of current. It is not clear, however, why this pin should burn only just below the thread, while other pins of the same kind on the same line were charred all the way down from the thread to the cross-arm.

Another curious result noticed in some pins on this same line is the softening of the threads so that they can be rubbed off with the fingers.

RELATION OF PINS AND INSULATORS.

Location of Line.	Voltage of Line.	Diameter of Insulator.	Height of Insulator.	Length of Pin Covered by Insulator.
		Inches.	Inches.	Inches.
Electra to San Francisco	60,000	11	11 ¼	12
Colgate to Oakland	60,000	11	11 ¼	8
Cañon Ferry to Butte	50,000	9	12	10 ½
Shawinigan Falls to Montreal	50,000	10	13	10 ¼
Santa Ana River to Los Angeles	33,000	6 ¾	4 ⅞	2 ½
Provo around Utah Lake	40,000	7	5 ¾	4 ¾
Spier Falls to Schenectady	30,000	8 ½	6 ¾	5 ¼
Niagara Falls to Buffalo	22,000	7 ½	7	5

The softened wood of the threads is not charred, but is said to have a sour taste and to resemble digested wood pulp. While the threads of a wooden pin are destroyed in this way the remainder of the pin may still remain perfect and show no charring.

RELATIONS OF PINS AND INSULATORS.

Location of Line.	Length of Pin Between Insulator and Cross-arm.	Distance from Outer Petticoat to Pin Through Air.	Distance from Lowest Petticoat to Pin Through Air.
	Inches.	Inches.	Inches.
Electra to San Francisco	0	10 ½	3 ½
Colgate to Oakland	3 ½	10	2 ½
Cañon Ferry to Butte	1 ½	0	1 ½
Shawinigan Falls to Montreal	3 ¼	9 ½	1
Santa Ana River to Los Angeles	3 ½	2 ¾	..
Provo around Utah Lake	3 ½	2 ½	..
Spier Falls to Schenectady	4	4	⅝
Niagara Falls to Buffalo	3	4 ½	2

In explanation of this disintegration of the threads of wooden pins it was stated that a number of these pins, the tops of which were reduced to a white powder, had been taken from the line between Niagara Falls and Buffalo, on which the voltage is 22,000, and that this powder proved on analysis to be a nitrate salt. This salt was thought to be the result of the action of nitric acid on the wood, it being supposed that the acid was formed by a static discharge acting on the oxygen and nitrogen of the air between the threads of the insulator and pin. In support of this view it was stated that an experimental line of galvanized-iron wire at Niagara Falls, which was operated at 75,000 volts continuously during nearly four months, turned black over its entire length of about two miles. This surface disintegration was not due to the normal action of the air, for similar wire at the same place remained bright when not used as an electrical conductor.

These facts seemed to indicate that the brush discharge from the wires carrying the 75,000-volt current developed nitric acid from the oxygen and nitrogen of the air, and that this acid attacked the wire.

One of the above-mentioned pins used on the Electra line was much charred and burned away at a point a little below the threads. The charred path of the current could also be traced down the side of the pin to the cross-arm, but this path was not as badly burned as the spot near the top of the pin.

A composite pin from a 33,000-volt line, probably a part of the transmission system between the Santa Ana River and Los Angeles, was burned through its wooden threads to the central iron bolt, along a narrow strip at one side. Every pin burned on this line was said to show the effects of the current in the way just described, but no cross-arms were burned and very few insulators punctured.

The composite pin was made up of a central iron bolt 10⅝ inches long, ½-inch in diameter, and with a thin head above the wooden threads, a sleeve of wood 2⅝ inches long and 1 inch in diameter in its threaded portion, and a sleeve of porcelain 3⅛ inches long and 1¼ inches in diameter at its upper and 2$^{11}/_{16}$ inches at its lower end. The sleeves of wood and porcelain were slipped over the central iron bolt so that the portions of the pin above the cross-arm measured 5⅞ inches. In this case the path of the leakage current seems to have been over both the exterior and interior surface of the insulator and then through the wooden sleeve to the central bolt and the cross-arm.

The facts just outlined certainly indicate a serious menace to the permanence and reliability of long, high-voltage transmission lines supported by insulators on wooden pins. If such results have been encountered on the lines above named, where some of the largest and best designs of insulators are employed, it is only fair to assume that similar destructive effects of leakage currents are taking place on many other lines that operate at high voltages.

It seems at least doubtful whether any enlargement or improvement of the insulators themselves will entirely avoid the destruction of their wooden pins in one of the ways mentioned. It is probable, but not certain, that further extension of distances through air and over insulator surfaces, both exterior and interior, between line wires and wooden pins, will prevent charring and burning of the latter by leakage currents. Much has already been done in the way of covering most of the pin above its cross-arm with the insulator parts, but even those portions of the pin that are best protected in this way are not free from burning.

Thus, on the Colgate line, eight inches of each pin is protected by the interior surface of its insulator, but these pins were charred quite as badly where best protected, up close to the thread, as they were down near the cross-arm. The same is true of the Electra line, where a porcelain sleeve runs up about the pin from the cross-arm to a point above the inner petticoat of each insulator, so that the entire length of the pin above the cross-arm is protected. On the Cañon Ferry line, a glass sleeve that virtually forms a part of each insulator, though mechanically separate from it, protects the pin from its threaded portion to within 1.5 inches of the cross-arm.

Insulators on the line from Shawinigan Falls to Montreal are each 13 inches long and extend down over the pin to within 1.5 inches of the cross-arm. The burned portion of each pin from the Santa Ana line was that carrying the threads, and thus in actual contact with that part of the insulator which was separated by the greatest surface distance from the line wire.

Aside from the burning of pins is the destruction of their threaded parts by some chemical agency that is developed inside of the tops of the insulators, as shown in the cases of the Colgate and Niagara lines. It does not appear that any improvement of insulators will necessarily prevent chemical action.

Though it may not be practicable to so increase the surface resistance of each insulator that the burning of wooden pins by leakage current will be prevented, the substitution of a conducting for an insulating pin may remedy the trouble. As the insulators, pins, and cross-arm form a path for the leakage current from wire to wire, the wooden pins by their resistance, especially when dry, must develop heat. In pins of steel or iron this heat would be trifling and would do no damage. With pins of good conducting material, like iron, the amount of leakage from wire to wire, with a given design of insulator, would, no doubt, be somewhat greater than the leakage with wooden pins.

It will be cheaper, however, to increase the resistance of new insulators up to the combined resistance of present insulators and their wooden pins than it will be to replace these pins when they are burned.

From all the evidence at hand, it seems that insulators which reduce the leakage of current over their surfaces to permissible limits as far as mere loss of energy is concerned, even with iron pins, will not prevent the charring and destruction of wooden pins.

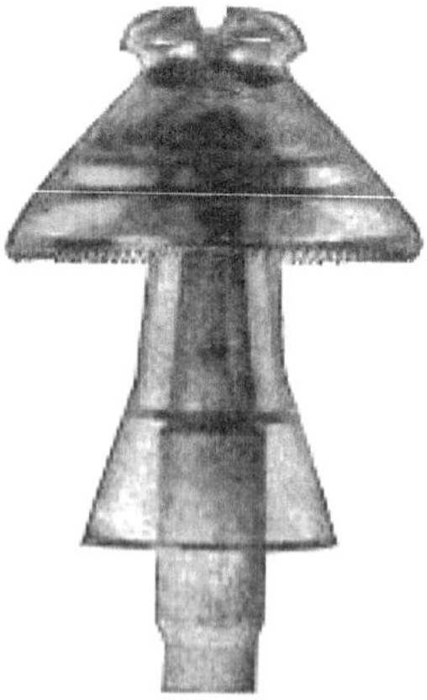

FIG. 90.—Glass Insulator and Sleeve on 50,000-volt Line Between Cañon Ferry and Butte, Mont.

When any suitable insulator is dry and clean it offers all necessary resistance to the leakage of current over its surface, and any resistance in the pin that carries the insulator is of small importance. If the resistance of an insulator needs to be reinforced by that of its pin in any case, it is when the surface of the insulator is wet or dirty. Unfortunately, however, the same weather conditions that deposit dirt or moisture on an insulator make similar deposits on its pin, and the resistance of the pin is lowered much more than that of the insulator by such deposits. The increase of current leakage over the surface of an insulator during rains and fogs usually does no damage to the insulator itself, but such leakage over the wet pin soon develops a surface layer of carbon that continues to act as a good conductor after the moisture that temporarily lowered the resistance has gone. Reasons like these have led some engineers to prefer iron pins with insulators that offer all of the resistance necessary for the voltage employed on the line.

It may be suggested that the use of iron pins will transfer the charring and burning to the wooden cross-arms, but this does not seem to be a necessary result. The comparative freedom of cross-arms from charring and burning where wooden pins are used seems to be due to the larger surface and lower resistance of the cross-arms. With iron pins having a shank of small diameter, so that the area of contact surface between the pin and the wooden cross-arm is relatively small, there may be some charring of the wood at this contact surface. Should it be thought desirable to guard against any trouble of this sort, the surface of the iron pin in contact with the cross-arm may be made ample by the use of large washers, or by giving each pin a greater diameter at the shank than elsewhere.

It may be noted that the pins with a central iron bolt only half an inch in diameter, that were used on the 33,000-volt Santa Ana line, were said to have caused no burning of their cross-arms in those cases in which the wooden threads about the top of the central bolt were burned through.

Another possible trouble with iron pins is that they, by their greater rate of expansion than glass or porcelain, will break their insulators. Such results can readily be avoided by cementing each iron pin into its insulator, instead of screwing the insulator onto the pin. Iron pins will, no doubt, cost somewhat more than those of wood, but this cost will in any event be only a small percentage of the total investment in a transmission line. Considering the cost of the renewals of wooden pins, there seems little doubt that on a line where the voltage and other conditions are such as to result in frequent burning, iron pins would be cheaper in the end.

Iron pins have already been adopted on a number of high-voltage lines. Not only iron pins, but even iron cross-arms and iron poles are in use on a number of transmission lines. On a long line now under construction in

Mexico, iron towers, placed as much as 400 feet apart, are used instead of wooden poles, and both the pins and cross-arms are also of iron. The 75-mile line from Niagara Falls to Toronto is carried entirely on steel towers.

The Vancouver Power Company, Vancouver, British Columbia, use a pin that consists of a steel bolt about 12 inches long fitted with a sleeve of cast iron 4½ inches long to enter the cross-arm, and a lead thread to screw into the insulator. On the 111-mile line of the Washington Power Company, of Spokane, which was designed to operate at 60,000 volts and runs to the Standard and Hecla mines, a pin consisting of a steel bar 1⅛ inches in diameter, with a cast-iron shank 2¹⁄₁₆ inches in diameter to enter the cross-arm, and with the lead threads for the insulator, is used.

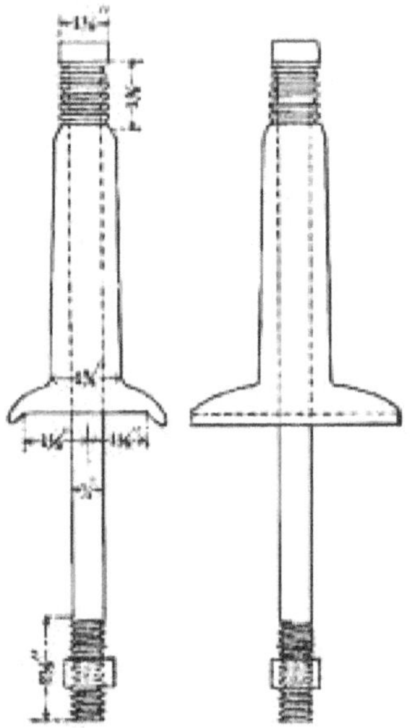

FIG. 92.—Iron Pins on Spier Falls Line.

On the network of transmission lines between Spier Falls, Schenectady, Albany, and Troy, in the State of New York, the insulators are supported on iron pins of two types. One of these pins, used at corners and where the strain on the wire line is exceptionally heavy, is made up of a wrought-iron bolt ¾-inch in diameter and 16½ inches long over the head, and of a malleable iron casting 8¾ inches long. This casting has a flange of 5 by 3¾ inches at its lower end that rests on the top of the cross-arm, and the bolt passes from the top of the casting down through it and the cross-arm. Threads are cut on the lower end of the bolt, and a nut and washer secure it in the cross-arm. The total height of this pin above the cross-arm is 9¼ inches.

For straight work on this line a pin with stem entirely of malleable iron, and a bolt that comes up through the cross-arm and enters the base of the casting, is used. The cast top of this pin has four vertical webs, and its rectangular base, which rests on the top of the cross-arm, is 3½ by 4 inches. The bolt that comes up through the cross-arm and taps into the base of the casting is ¾-inch in diameter. The cast part of this pin has such a length that the top of its insulator is carried 10¾ inches above the cross-arm. For the casting the length is 9¼ inches.

Both of the types of iron pins in use on the Spier Falls lines are secured to their insulators with Portland cement poured into the pin hole while liquid when the insulator is upside down and the pin is held centrally in its hole. The top of each casting is smaller in diameter than the hole in the insulator, and is grooved so as to hold the cement.

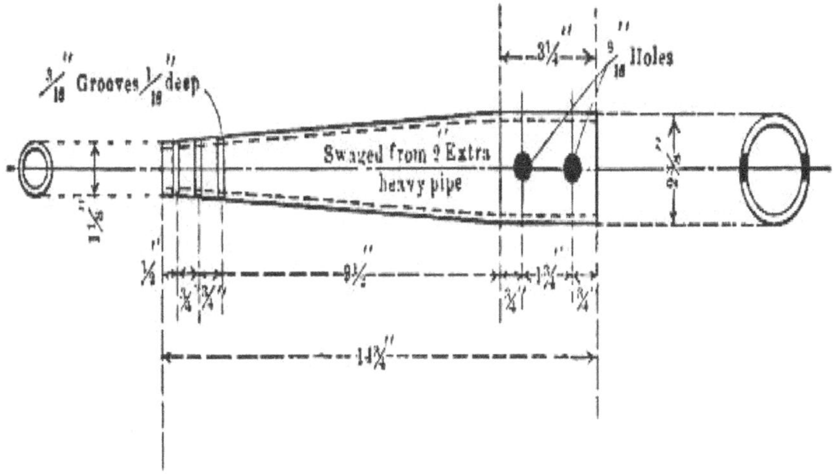

FIG. 93.—Standard Pin, Toronto and Niagara Line.

On a long line designed for 60,000 volts, and recently completed in California, wooden pins are used with porcelain insulators, each 14 inches in diameter and 12½ inches high. Each of these pins is entirely covered with sheet zinc from the cross-arm to the threaded end, and it is expected that this metal covering will protect the wood of the pin from injury by the leakage current.

CHAPTER XXI.
INSULATORS FOR TRANSMISSION LINES.

Line insulators, pins, and cross-arms all go to make up paths of more or less conductivity between the wires of a transmission circuit. The amount of current flowing along these paths from one conductor to another in any case will depend on the combined resistance of the insulators, pins, and cross-arm at each pole.

As a general rule, the wires of high-voltage transmission circuits are used bare because continuous coverings would add materially to the cost with only a trifling increase in effective insulation against high voltages. In some instances the wires of high-pressure transmission lines have individual coverings for short distances where they enter cities, but often this is not the case. At Manchester, N. H., bare conductors from water-power plants enter the sub-station, well within the city limits, at 12,000 volts. From the water-power at Chambly the bare 25,000-volt circuits, after crossing the St. Lawrence River over the great Victoria bridge, pass overhead to a terminal-house near the water-front in Montreal. In order to reach the General Electric Works, the 30,000-volt circuits from Spier Falls enter the city limits of Schenectady, N. Y., with bare overhead conductors.

Where transmission lines pass over a territory exposed to corrosive gases, it is sometimes desirable to give each wire a weather-proof covering. An instance of this sort occurs near Niagara Falls where the aluminum conductors forming one of the circuits to Buffalo are covered with a braid that is saturated with asphaltum for some distance.

Each path, formed by the surface of the insulators of a line and the pins and cross-arm by which they are supported, not only wastes the energy represented by the leakage current passing over it, but may lead to the charring and burning of the pins and cross-arm by this current. To prevent such burning, the main reliance is to be placed in the surface resistance of the insulators rather than that of pins and cross-arms. These insulators should be made of glass or porcelain, and should be used dry—that is, without oil. In some of the early transmission lines, insulators were used on which the lower edges were turned inward and upward so that a circular trough was formed beneath the body of the insulator, and this trough was filled with heavy petroleum. It was found, however, that this trough of oil served to collect dirt and thus tended to lower the insulation between wire and cross-arm, so that the practice was soon abandoned. Glass and porcelain insulators are rivals for use on high-tension lines, and each has advantages of its own. Porcelain insulators are much stronger mechanically than are those of glass, and are not liable to crack because of unequal internal expansion, a result sometimes met with where glass insulators are exposed to a hot morning sun. In favor of glass insulators it may be said that their insulating properties are quite uniform, and that, unlike porcelain, their internal defects are often apparent on inspection. In order to avoid internal defects in large porcelain insulators, it has been found necessary to manufacture some designs in several parts and then cement the parts of each insulator together.

Defective insulators may be divided into two classes—those that the line voltage will puncture and break and those that permit an excessive amount of current to pass over their surfaces to the pins and cross-arms. Where an insulator is punctured and broken, the pin, cross-arm, and pole to which it is attached are liable to be burned up. If the leakage of current over the surface of an insulator is large, not only may the loss of energy on the line where the insulator is used be serious, but this energy follows the pins and cross-arm in its path from wire to wire, and gradually chars the former, or both, so that they are ultimately set on fire or break through lack of mechanical strength. The discharge over the surface of an insulator may be so large in amount as to have a disruptive character, and thus to be readily visible. More frequently this surface leakage of current over insulators is of the invisible and silent sort that nevertheless may be sufficient in amount to char, weaken, and even ultimately set fire to pins and cross-arms.

All insulators, whether made of glass or porcelain, should be tested electrically to determine their ability to resist puncture, and to hold back the surface leakage of current, before they are put into practical use on high-tension lines. Experience has shown that inspection alone cannot be depended on to detect defective glass insulators. Electrical testing of insulators serves well to determine the voltage to which they may be subjected in practical service with little danger of puncture by the disruptive passage of current through their substance. It is also possible to determine the voltage that will cause a disruptive discharge of current over the surface of an insulator, when the outer part of this surface is either wet or dry. This is as far as electrical tests are usually carried, but it seems desirable that such tests should also determine the amount of silent, invisible leakage over the surface of insulators both when they are wet and when they are dry, at the voltage which their circuits are intended to carry. Such a test of silent leakage is important because this sort of leakage chars and weakens insulator pins, and sets fire to them and cross-arms, besides representing a waste of energy.

The voltage employed to test insulators should vary in amount according to the purpose for which any particular test is made. Glass and porcelain, like many other solid insulators, will withstand a voltage during a few minutes that will cause a puncture if continued indefinitely. In this respect these insulators are unlike air, which allows a disruptive discharge at once when the voltage to which it is exposed reaches an amount that the air cannot permanently withstand. Because of this property of glass and porcelain insulators, it is necessary in making a puncture test to employ a voltage much higher than that to which they are to be permanently exposed. In good practice it is thought desirable to test insulators for puncture with at least twice the voltage of the circuits which they will be required to permanently support on transmission lines.

For the first transmission line from Niagara Falls to Buffalo, which was designed to operate at 11,000 volts, the porcelain insulators were tested for puncture with a voltage of 40,000, or nearly four times that of the circuits they were to support.

Porcelain insulators for the second line between Niagara Falls and Buffalo, after the voltage of transmission had been raised to 22,000, were given a puncture test at 60,000 volts. Of these insulators tested at 60,000 volts only about three per cent proved to be defective. These puncture tests were carried out by placing each insulator upside down in an open pan containing salt water to a depth of two inches, partly filling the pin hole of the insulator with salt water, and then connecting one terminal of the testing circuit with a rod of metal in the pin hole, and the other terminal with the pan. Alternating current was employed in these tests, as is usually the case (Volume xviii., Transactions A. I. E. E., pp. 514 to 520). For the transmission lines between Spier Falls, Schenectady, Albany, and Troy, where the voltage is 30,000, the insulators were required to withstand a puncturing test with 75,000 volts for a period of five minutes after they had been soaked in water for twenty-four hours.

There is some difference of opinion as to the proper duration of a puncturing test, the practice in some cases being to continue the test for only one minute on each insulator, while in other cases the time runs up to five minutes or more. As a rule, the higher the testing voltage compared with that under which the insulators will be regularly used, the shorter should be the period of test. Instead of being tested in salt water as above described, an insulator may be screwed onto an iron pin of a size that fits its threads, and then one side of the testing circuit put in contact with the pin and the other side connected with the wire groove of the insulator. Care should be taken where an iron pin is used either in testing or for regular line work, that the pin is not screwed hard up against the top of the insulator, as this tends to crack off the top, especially when the pin and insulator are raised in temperature. Iron expands at a much higher rate than glass or porcelain, and it is desirable to cement iron pins into insulators rather than to screw them in. There seems to be some reason to think that an insulator will puncture more readily when it is exposed to severe mechanical stress by the expansion of the iron pin on which it is mounted.

Tests of insulators are usually made with alternating current, and the form of the voltage curve is important, especially where the test is made to determine what voltage will arc over the surface of the insulator from the line wire to the pin. The square root of the mean square for two curves of alternating voltage or mean effective voltage, as read by a voltmeter, may be the same though the maximum voltages of the two curves differ widely. In tests for the puncture of insulators, the average alternating voltage applied is more important than the maximum voltage shown by the highest points of the pressure curve, because of the influence of the time element with glass and porcelain. On the other hand, when the test is to determine the voltage at which current will arc over the insulator surface from the line wire to the pin, the maximum value of the pressure curve should be taken into consideration because air has no time element, but permits a disruptive discharge under a merely instantaneous voltage.

Alternators used in transmission systems usually conform approximately to a sine curve in the instantaneous values of the pressures they develop, and it is therefore desirable that tests on line insulators be made with voltages whose values follow the sine curve. Either a single transformer or several transformers in series may be employed to step up to the required voltage, but a single transformer will usually give better regulation and greater accuracy. An air-gap between needle points is not a very satisfactory means by which to determine the average voltage on a testing circuit, because, as already pointed out, the sparking distance between the needle points depends mainly on the maximum instantaneous values of the voltage, which may vary with the load on the generator, and the saturation of its magnets. For accurate results a step-down voltmeter transformer should be used on the testing circuit.

An insulator that resists a puncture test may fail badly when subjected to a test as to the voltage that will arc over its surface from line wire to pin. This arc-over test should be made with the outer surface of the insulator both wet and dry. For the purpose of this test the insulator should be screwed onto an iron pin, or onto a wooden pin that has been covered with tinfoil. One wire of the testing circuit should then be secured in the groove of the insulator, and the other wire should be connected to the iron or tin foil of the pin. The voltage that will arc over the surface of an insulator from the line wire to the pin depends on the conditions of that surface and of the air. In light air, such as is found at great elevations, an arc will jump a greater distance than in dry air near the sea-level. A fog increases the distance that a given voltage will jump between a line wire and its insulator pin, and a heavy rain lengthens the distance still further. The heavier the downpour of rain the greater is the distance over the outside surface of an insulator that a given voltage will arc over. The angle at which the falling water strikes the insulator surface also has an influence on the voltage required to arc over that surface, a deviation from a downpour perpendicular to the plane of the lower edge of the petticoat of the insulator seeming to increase the arcing distance for a given voltage.

An insulator should be given an arc-over test under conditions that are approximately the most severe to be met in practice. These conditions can perhaps be fairly represented by a downpour of water that amounts to a depth of one inch in five minutes for each square inch of the plane included by the edge of the largest petticoat of the insulator, when the direction of the falling water makes an angle of forty-five degrees with that plane. A precipitation of one inch in depth on a horizontal plane during five minutes seems to be a little greater than any recorded by the United States Weather Bureau. Under the severe conditions just named, the voltage required to arc over the insulator surface from line wire to pin should be somewhat greater at least than the normal voltage of the circuit where the insulator is to be used. For the transmission line between Spier Falls and Schenectady, on which the maximum voltage is 30,000, the insulators were required to stand a test of 42,000 volts when wet, without arcing over from line wire to pin. In these wet tests the water should be sprayed evenly onto the insulator surface like rain, and the quantity of water that strikes the insulator in a given time should be measured.

When the outside of an insulator is wet with rain, it is evident that most of the resistance between the line wire and the insulator pin must be offered by the inside surface of the petticoat of the insulator. For this reason an insulator that is to withstand a very high voltage so that no arc will be formed over its wet outside surface must have a wide, dry surface under its petticoat. In some tests of line insulators reported in Volume xxi., Transactions A. I. E. E., p. 314, the results show that the voltage required to arc over from line wire to pin depends on the shortest distance between them, rather than on the distance over the insulator surface. Three insulators, numbered 4, 5, and 7 in the trial, were in each case tested by a gradual increase of voltage until a discharge took place between the wire and pin. The pins were coated with tinfoil, and the testing voltage was applied to the tie wire on each insulator and to the tinfoil of its pin. Insulators 4, 5, and 7 permitted arcs from wire to pin when exposed to 73,800, 74,700, and 74,700 volts respectively, the surfaces of all being dry and clean. The shortest distances between wires and pins over insulator surface and through air were $6\frac{5}{8}$, $6\frac{1}{4}$, and $7\frac{7}{8}$ inches respectively for the three insulators, so that the arcing voltages amounted to 11,140, 11,952, and 9,479 per inch of these distances. Measured along their surfaces, the distances between wires and pins on these three insulators were 8, $11\frac{1}{4}$, and $15\frac{1}{2}$ inches respectively, so that the three arcing voltages, which were nearly equal, amounted to 9,225, 6,640, and 4,819 per inch of these distances. These figures make it plain that the arcing voltage for each insulator depends on the shortest distance over its surface and through the air, from wire to pin. It might be expected that the voltage in any case would arc equal distances over clean, dry insulator surface or through the air, and the experiments just named indicate that this view is approximately correct. The sparking distance through air between needle points, which is greater than that between smooth surfaces, is 5.85 inches with 70,000 volts, and 7.1 inches with 80,000 volts according to the report in Volume xix., A. I. E. E., p. 721. Comparing these distances with the shortest distances between wires and pins in the tests of insulators numbered 4, 5, and 7, which broke down at 73,800 to 74,700 volts when dry, it seems that a given voltage will arc somewhat further over clean, dry insulator surface than it will through air. This view finds support from the fact that only a part of each of the shortest distances between wire and pin was over insulator surface, the remainder being through air alone.

The fact that the dry part of the surface of an insulator and the air between its lower wet edge and the pin or cross-arm offer most of the resistance between the line wire and the pin and cross-arm is plainly brought out by the results of the tests above mentioned, in the cases of insulators numbered 4 and 7. While 73,800 volts were required to arc from line-wire to pin when the entire insulator was dry and clean, the arc was formed at only 53,400 volts during a moderate rain-storm, in the case of No. 4 insulator. With insulator No. 7 the arcing voltage was 74,700 when the entire surface was clean and dry, but the arc from wire to pin was started at 52,800 volts during a moderate rain. No. 5 insulator seems to present an erratic result, for when dry and clean the arc jumped from wire to pin at 74,700 volts, and yet during a moderate rain no arc was formed until a voltage of 70,400 was reached. For each of the seven insulators on which tests are reported as above, the voltage required to arc from line wire to pin was nearly or quite as great during a dry snow-storm as when the insulator surface was clean and dry. When the insulators were covered with wet snow their surface insulation broke down at voltages that were within ten per cent above or below the arcing voltages during a moderate rain in five cases. With two insulators the arcing voltages, when they were covered with wet snow, were only about sixty per cent of the voltages necessary to break down the surface insulation between wire and pin during a moderate rain.

When the outside surface of an insulator is wet, as during a moderate rain, it seems that the under surface of the insulator, and the distance through air from the lower wet edge of the insulator to the pin or cross-arm, make up most of the insulation that prevents arcing over from the wire to the pin or cross-arm. It further appears that it is useless to extend the distance across the dry under surface of the insulator indefinitely without a corresponding increase of the direct distance through air from the lower wet edge of the insulator to the wood of cross-arm or pin. Insulator No. 7 in the tests under consideration had a diameter at the lower edge of its outer petticoat of seven inches, and was mounted on a standard wooden pin. The diameter of this pin in the plane of the lower edge of the insulator was probably about 1¼ inches, so that the radial distance through air from this edge to the pin must have been 2⅞ inches approximately. During a moderate rain the surface insulation of this insulator broke down and an arc was formed from wire to pin with 52,800 volts. The sparking distance between needle points at 50,000 volts is 3.55 inches, according to Volume xix., A. I. E. E., p. 721, and must be shorter between smooth surfaces, such as the wire and pin in question, so that nearly all of the 52,800 volts in this case must have been required to jump the 2⅞ inches of air, leaving very little to overcome the slight resistance of the wet outside surface of the insulator. On this insulator the surface distance from wire to pin was 15-½ inches, while the shortest breaking distance was only 7⅞ inches, so that the distance across the dry under surface of the insulator must have been 15½ - (7⅞ - 2⅞) = 10½ inches approximately. It is evidently futile to put a path 10½ inches long across dry insulator surface in parallel with a path only 2⅞ inches long in air, as an arc will certainly jump this shorter path long before one will be formed over the longer. The same line of reasoning applies to No. 3 insulator in this test, which had a diameter of 6¾ inches, a surface distance from wire to pin of 13 inches, and a minimum distance of 7¼ inches, and whose surface insulation broke down at 48,600 volts during a moderate rain. The necessity of increasing the distance between the lower wet edges of insulators and the pins and cross-arm, as well as the distance across the dry under surfaces of insulators, led to the adoption of the so-called umbrella type for some high-voltage lines. In this type of insulator the main or outer petticoat is given a relatively great diameter, and instead of being bell-shaped is only moderately concave on its under side. With an insulator of this type mounted on a large, long pin, the lower edge of the umbrella-like petticoat may be far removed from the pin and cross-arm. Beneath the large petticoat of such insulators for high voltages there are usually one or more smaller petticoats or sleeves that run down the pin, and increase the distance between it and the lower edge of the largest petticoat.

INSULATORS ON TRANSMISSION LINES.

Location of Line.	Voltage of Line.	Material of Insulator.	Inches Diameter of Insulator.	Inches Height of Insulator.
Electra to San Francisco	60,000	Porcelain	11	11 1/4
Colgate to Oakland	60,000	Porcelain	11	11 1/4
Cañon Ferry to Butte	50,000	Glass	9	12
Shawinigan Falls to Montreal	50,000	Porcelain	10	13 1/2
Provo around Utah Lake	40,000	Glass	7	5 3/4
Santa Ana River to Los Angeles	33,000	Porcelain	6 3/4	4 7/8
Spier Falls to Schenectady	30,000	Porcelain	8 1/2	6 3/4
Apple River Falls to St Paul	25,000	Glass	7	5 3/4
Chambly to Montreal	25,000	Porcelain	5 1/2	6 1/2
Niagara Falls to Buffalo	22,000	Porcelain	7 1/2	7
Portsmouth to Pelham, N. H.	13,000	Porcelain	5 1/4	3 3/4
Garvins Falls to Manchester, N. H.	12,000	Glass	5	4 3/4

The inner petticoat or sleeve that runs down over the pin and sometimes reaches nearly to the cross-arm, of course becomes wet on its outside surface and at its lower edge during a rain; but between this lower wet part of the inner petticoat, or sleeve, and the lower wet edge of the larger outside petticoat, there is a wide, dry strip of insulator surface. A result is that an arc over the surface of the outside petticoat can reach the wet edge of the sleeve only by crossing the strip of dry under surface or jumping through the air.

The same type of insulator is used on the 60,000-volt lines between Electra and San Francisco and between Colgate and Oakland, each insulator having an outer petticoat 11 inches in diameter and one inner petticoat or sleeve 6½ inches in diameter. This inner petticoat runs down the pin for a distance of 7½ inches below the outer petticoat. Slightly different pins are used for mounting the insulators on the two transmission lines just named, so that on the former the distance through air from the lower edge of the outer petticoat to the cross-arm is 11 inches, and on the latter the corresponding distance is 11½ inches. On the Electra line the lower edge of the inner petticoat of each insulator is about 3½ inches, and on the Colgate line about 4 inches above the cross-arm.

Location of Line.	Inches from Top of Insulator to Cross-arm.	Inches from Outside Petticoat to Cross-arm.	Inches from Lowest Petticoat to Cross-arm.	Inches from Edge of Outside to Edge of Lowest Petticoat.
Electra to San Francisco	14 ½	11	3 ½	7 ½
Colgate to Oakland	15	11 ½	4	7 ½
Cañon Ferry to Butte	13 ½	7 ¾	1 ½	6 ¼
Shawinigan Falls to Montreal	16 ¼	11 ¾	3 ¼	8 ½
Santa Ana River to Los Angeles	8 ⅝	3 ¾	3 ¾	0
Spier Falls to Schenectady	10 ¾	7 ⅜	4 ¼	3 ⅜
Niagara Falls to Buffalo	10	5 ½	3	2 ½
Chambly to Montreal	8 ½	4 ½	2	2 ½

On each of the lines named in this table the wires are strung on the tops of their insulators.

The Cañon Ferry line is carried on insulators each of which has three short petticoats and a long separate sleeve that runs down over the pin to within 1½ inches of the cross-arm. This sleeve makes contact with its insulator near the pin hole. The outside petticoat of each insulator on this line is 7¾ inches above the cross-arm and 6¼ inches above the lower end of the sleeve. Both the main insulator and the sleeve, in this case, are of glass.

White porcelain insulators are used to support the 50,000-volt Shawinigan line, and are of a recent design. Each of these insulators has three petticoats ranged about a central stem so that their lower edges are 4½ inches, 9 inches, and 13 inches respectively, below the top. The highest petticoat is 10 inches, the intermediate 9¾ inches, and the lowest 4¼ inches in diameter. The height of this insulator is 13 inches, compared with 11¼ inches for those used on the Electra and Colgate lines and 12 inches for the combined insulator and sleeve used on the Cañon Ferry line. When mounted on its pin, this insulator on the Shawinigan line holds its wire 16¼ inches above the cross-arm, compared with a corresponding distance of 14½ inches on the Electra, 15 inches on the Colgate, and 13½ inches on the Cañon Ferry line. The two upper petticoats on each of these insulators are much less concave than the lowest one, and the edges of all three stand respectively 11¾, 7¼, and 3¼ inches above the cross-arm. From the edge of the top to the edge of the bottom petticoat the direct distance is 8½ inches.

Of the three transmission lines above named that operate at 50,000 to 60,000 volts, that between Shawinigan Falls and Montreal leads as to distances between the line wire and insulator petticoats and the cross-arm. On the Santa Ana line, where the voltage is 33,000, the insulator is of a more ordinary type, being of porcelain, 6¾ inches in diameter, 4⅞ inches high, and having the lower edges of its three petticoats in the same plane. Each of these insulators holds its wire 8⅝ inches above the cross-arm, and has all of its petticoats 3½ inches above the cross-arm. Unlike the three insulators just described, which are mounted on wooden pins, this Santa Ana insulator has a pin with an iron core, wooden threads, and porcelain base. This base extends up from the cross-arm a distance of 3⅛ inches, and the wooden sleeve, in which the threads for the insulator are cut, runs down over the central bolt of the pin to the top of the porcelain base, which is ⅝-inch below the petticoats.

The 30,000-volt lines from Spier Falls are carried 10¾ inches above their cross-arms by triple petticoat porcelain insulators. Each of these insulators is 8½ inches in diameter, 6¾ inches high, and is built up of three parts cemented together. A malleable-iron pin cemented into each insulator with pure Portland cement carries the outside petticoat 7½ inches and its lowest petticoat 4¼ inches above the cross-arm. When the voltage on the Spier Falls lines was raised from about 13,000 to 30,000, the circuits being carried in part by one-piece porcelain insulators, a number of these insulators were punctured at the higher pressures, and some cross-arms and poles were burned as a result. No failures resulted on those parts of these lines where the three-part insulators were in use. The second pole line between Niagara Falls and Buffalo was designed to carry circuits at 22,000 volts, or twice that for which the first line was built. Porcelain insulators were employed on both of these lines, but while the 11,000-volt line was carried on three-petticoat insulators, each with a diameter of 7 inches and a height of 5½ inches, the 22,000-volt line was mounted on insulators each 7½ inches in diameter and 7 inches high, with only two petticoats. The older insulator has its petticoats 2 inches above the cross-arm, and the lower petticoat of the new insulator is 3 inches above the arm. These two insulators illustrate the tendency to lengthen out along the insulator axis as the voltage of the circuits to be carried increases.

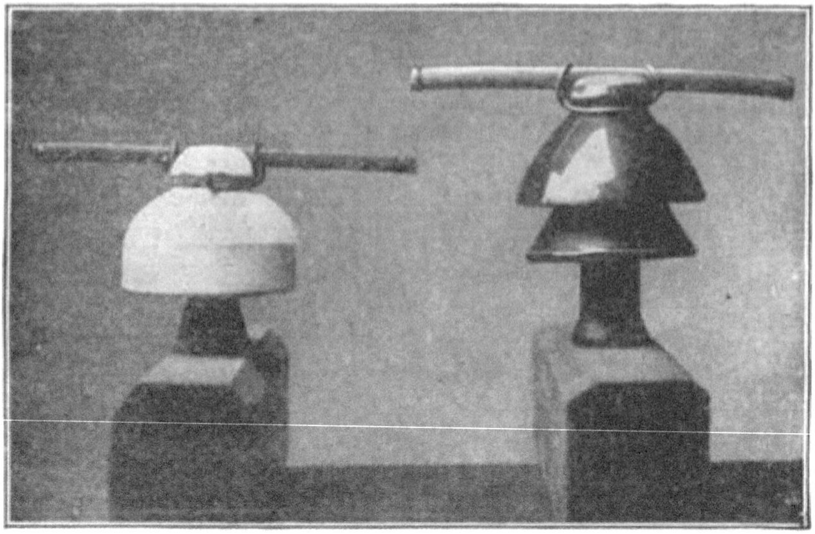

FIG. 93A.—The Old and New Insulators on the Niagara Falls-Buffalo Line.

For future work at still higher voltages, the advantage as to both first cost and insulating qualities seems to lie with insulators that are very long in an axial direction, and which have their petticoats arranged one below the other and all of about the same diameter, rather than with insulators of the umbrella type, like those on the Electra and Colgate lines.

CHAPTER XXII.
DESIGN OF INSULATOR PINS FOR TRANSMISSION LINES.

Bending strains due to the weights, degree of tension, and the directions of line wires, plus those resulting from wind-pressure, are the chief causes that lead to the mechanical failure of insulator pins.

Considering the unbalanced component of these forces at right angles to the axis of the pin, which alone produce bending, each pin may be considered as a beam of circular cross section secured at one end and loaded at the other.

For this purpose the secured end of the beam is to be taken as the point where the pin enters its cross-arm, and the loaded end of the beam is the point where the line wire is attached to the insulator. The distance between these two points is the length of the beam. The maximum strain in the outside fibres of a pin measured in pounds per square inch of its cross section, represented by S, may be found from the formula,

$$S = P X .0982 D^3$$

where P is the pull of the wire in pounds, D is the diameter of the pin at any point, and X is the distance in inches of that point from the wire. Inspection of this formula shows that S, the maximum strain at any point in the fibres of a pin, when the pull of the line-wire, P, is constant, increases directly with the distance, X, from the wire to the point where the strain, S, takes place. This strain, S, with a constant pull of the line wire, decreases as the cube of the diameter, D, at the point on the pin where S occurs increases. That cross section of a pin just at the top of its hole in the cross-arm is thus subject to the greatest strain, if the pin is of uniform diameter, because this cross section is more distant from the line wire than any other that is exposed to the bending strain. For this reason it is not necessary to give a pin a uniform diameter above its cross-arm, and in practice it is always tapered toward its top. Notwithstanding this taper, the weakest point in pins as usually made is just at the top of the cross-arm, and it is at this cross section where pins usually break. This break comes just below the shoulder that is turned on each pin to prevent its slipping down through the hole in its cross-arm. If the shoulder on a pin made a tight fit all around down onto the cross-arm, the strength of the pin to resist bending would be thereby increased, but it is hard to be sure of making such fits, and they should not be relied on to increase the strength of pins. By giving a pin a suitable taper from its shoulder at the cross-arm to its top, the strain per square inch, S, in the outside fibres of the pin may be made constant for every cross section throughout its length above the cross-arm, whatever that length may be. The formula above given may be used to determine the diameters of a pin at various cross sections that will make the maximum stress, S, at each of these cross sections constant. By transposition the formula becomes

$$D^3 = P\ .0982\ S\ X.$$

Where the pin is tapered so that S is constant for all cross sections, then for any pull, P, of the line wire on the pin the quantity (P.0982 S) must be constant at every diameter, D, distant any number of inches, X, from the point where the wire is attached. If the constant, (P.0982 S) is found for any one cross section of a pin, therefore, the diameter at each other cross section with the same maximum stress, S, may be readily found by substituting the value of this constant in the formula. The so-called "standard" wooden pin that has been very generally used for ordinary distribution lines, and to some extent even on high-voltage transmission lines, has a diameter of nearly 1.5 inches just below the shoulder. The distance of the line wire above this shoulder varies between about 4.5 and 6 inches, according to the type of insulator used, and to whether the wire is tied at the side or top of the insulator. If the line wire is tied to the insulator 5 inches above the shoulder of one of the standard pins, then X becomes 5, and D becomes 1.5 in the formula last given. From that formula by transposition and substitution

$$P.0982\ S = D^3X = (1.5)^3 5 = 0.675.$$

Substituting 0.675 for the quantity P0.0982 S in the formula $D^3 = P0.0982\ S\ X$ gives the formula $D^3 = 0.675\ X$, from which the diameters at all cross sections of a tapered pin above its shoulder, that will give it a strength just equal to that of a section of 1.5 inches diameter and 5 inches from the line wire, may be found. To use the formula for this purpose it is only necessary to substitute any desired values of X therein and then solve in each case for the corresponding values of D. Let it be required, for instance, to determine what diameter a pin should have at a cross section one inch below the line wire in order that the maximum strain at that cross section may equal the corresponding strain at a cross section five inches below the line wire and of 1.5 inch diameter. Substituting one as the value of X, the last-named formula becomes $D^3 = 0.675$, and from this, $D = 0.877$, which shows that the diameter of the pin one inch below the line wire should be 0.877-inch. A similar calculation will show that if a pin is long enough so that a cross section above the cross-arm is 12 inches below the line wire, the diameter of this cross section should be equal to the cube root of $0.675 \times 12 = 8.1$, which is 2.008, or practically two inches. It should be observed that the calculations just made have nothing to do with the ability of a pin to resist any particular pull of its line wire. These calculations simply show what diameters a pin should have at different distances below its line wire in order that the maximum stress at each of its cross sections may equal that at a cross section 5 inches below the wire where the diameter is 1.5 inches. In Vol. xx., A. I. E.

E., pp. 415 to 419, specifications are proposed for standard insulator pins based on calculations like those just made. As a result of such calculations, the following table for the corresponding values of X and D, as used in the above formula, are there presented, each expressed in inches.

X	D	
1	0	.877
2	1	.106
3	1	.263
4	1	.395
5	1	.500
6	1	.592
7	1	.678
8	1	.754
9	1	.825
10	1	.888
11	1	.95
13	2	.06
15	2	.17
17	2	.25
19	2	.34
21	2	.42

A pin twenty-one inches long between the line wire and the cross-arm will have a uniform strength to resist the pull of the wire if it has the diameter given in this table at the corresponding distances below the line wire. From this it follows that a pin of any length between wire and cross-arm corresponding to X in the table will be equally strong to resist a pull of the line wire as a standard 1.5-inch diameter pin with its wire five inches above the cross-arm. In other words, if a pin that is twenty-one inches long between the line wire and the cross-arm has the diameters given in the table at the corresponding distances below the wire, then a pin of equal strength to resist bending, and of any shorter length, would correspond in the part above the cross-arm to an equal length cut from the top end of the longer pin.

Designating that part of a pin that is above the cross-arm as the "stem," and that part in the cross-arm as the "shank," each pin in the specifications under consideration is named by the length of its stem, as a 5-, 7- or 11-inch pin. It is proposed that each pin of whatever length be threaded for a distance of 2.5 inches at the top of its stem with four threads per inch, the sides of each thread being at an angle of ninety degrees with each other. Each thread is to cut into the pin about 3/32 inch, come to a sharp angle at the bottom, and be about 1/16 inch wide on top. At the end of the pin the proposed diameter over the thread is one inch in all cases, and at the lower end of the threaded portion the outside diameter is 1.25 inches. Near the end of the pin the diameter at the bottom of the thread is thus only 13/16 inch, and the corresponding diameter at the lower end of the threaded portion is about 1 1/16 inches on all pins. Each pin is to have a square shoulder to rest on the cross-arm, and the diameter of this shoulder is to be 3/8 inch greater than the nominal diameter of the shank of the pin. The proposed length of this shoulder on all pins is 1/4 inch before the taper begins. The actual diameter of the shank of each pin just below its shoulder is to be 1/32 inch less than the nominal diameter, and the actual diameter of the lower end of each shank is to be 1/16 inch less than the nominal diameter. With these explanations the proposed sizes of pins have dimensions as follows in inches:

Length of Stem.	Length of Shank.	Nominal Diameter of Shank.
5	4 1/4	1 1/2
7	4 1/4	1 3/4
9	4 1/4	1 7/8
11	4 3/4	2
13	4 3/4	2 1/8
15	4 3/4	2 1/4
17	5 3/4	2 3/8
19	5 3/4	2 1/2

In order rightly to appreciate the utility of this table of proposed standard pins, it is necessary to have in mind the fact that all the dimensions are based on the assumption that a wooden pin with a shank of one and one-half inches diameter, and with its line wire attached five inches above the cross-arm, is strong enough for general use on transmission lines. Such an assumption covers a wide range of practice, but its truth may well be doubted for many cases. That this assumption does form the basis of the entire table is clearly shown by the fact that the calculated diameter at the shank of each pin is made to depend on a uniform pull, *P*, of the line wire, giving a uniform maximum stress, *S*, in the outer fibres of the wood just where the shank joins the stem. In other words, every pin in the table is designed to break with a uniform pull of the line wire, provided that the point on the insulator where the wire is attached is just on a level with the top of its pin in each case. It will at once occur to practical men that while a five-inch pin with one and one-half inch shank, or a larger pin of equal ability to resist the pull of a line wire, may be strong enough for the conductors of some transmission lines, this same pin may be entirely too weak for the longer spans, sharper angles, and heavier conductors of other lines.

Thus, on the sixty-five-mile line between Cañon Ferry and Butte, Mont., each conductor is of copper and has a cross section of 106,500 cm., while on the older line between Niagara Falls and Buffalo each copper conductor has a cross section of 350,000 cm. Evidently with equal conditions as to length of span, amount of sag, and sharpness of angles on these two lines, pins ample in strength for the smaller wire might be much too weak for the larger wire.

A little consideration will show that it is neither rational nor desirable to adopt pins of uniform strength for all transmission lines, but that several degrees of strength are necessary to correspond with the range in sizes of conductors in regular use. The size of pins for use on any transmission line, when the maximum bending strain exerted by the conductors has been determined, should be found by calculation and experiment, or by experiment alone. According to Trautwine, the average compressive strength of yellow locust is 9,800 pounds, of hickory 8,000 pounds, and of white oak 7,000 pounds per square inch in the direction of the grain. These compressive strengths are less than the tensile strengths of the same woods, and should therefore be employed in calculation, since the fibres on one side of a bending pin are compressed while the fibres on the other side are elongated. Substituting 1,000 for the value of S in the formula, $S = P X .0982 D^3$, and also 5 for the value of X, and 1½ for the value of D, the resulting value of P is found to be 736.5 pounds. This result shows that with a locust pin of 1½ inches diameter at the shank, and with its line wire attached five inches above the shoulder, the unbalanced side pull of the wire that will break the pin by bending is 736 pounds, provided that the wood of the pin has a strength of 1,000 pounds per square inch in compression. As all of the proposed standard pins in the above table are designed for uniform strength to resist the same pull of a line wire attached on a level with the top of the pin in each case, it follows that the pull of 736 pounds by the wire will break any one of these pins under the conditions stated.

The calculation just made takes no account of the fact that the actual diameter of the shank of each pin just below the shoulder is $\frac{1}{32}$ inch less than the nominal diameter, but this of course reduces the strength somewhat. Trautwine states that the figures above given for the compressive strengths of wood are only averages and are subject to much variation. Of course no pin should be knowingly loaded in regular practice to the breaking point, and to provide against variations in the strength of wood, and for unexpected strains, a liberal factor of safety, say four, should be adopted in fixing the maximum strains on insulator pins. Applying this factor to the calculations just made, it appears that the maximum pull of the line wire at the top of any one of the above proposed standard pins should not exceed $736 \div 4 = 184$ pounds in regular work. A little calculation will readily show that the side pull of some of the larger conductors now in use on transmission lines will greatly exceed 184 pounds under conditions, as to sag, angles and wind pressure, that are frequently met in practice.

On page 448, Vol. xx., A. I. E. E., some tests are reported on six locust wood pins with shank diameters of $1\frac{7}{16}$ to $1\frac{1}{2}$ inches. Each of these pins was tested by inserting its shank in a hole of $1\frac{1}{2}$ inches diameter in a block of hard wood, and then applying a strain at about right angles to the pin and about $4\frac{1}{2}$ inches from the block by means of a Seller's machine. The pull on each pin was applied gradually, and in most of the pins the fibres of the wood began to part when the side pull reached 700 to 750 pounds, though the maximum loads sustained were about ten per cent above these figures. The average calculated value of S, the compressive strength of the wood in these pins, was 11,130 pounds per square inch on the basis of the loads at which the fibres of the wood began to break, and 13,623 pounds per square inch for the loads at which the pins gave way. On pages 650 to 653 of the volume last cited, results are reported of tests on twenty-two pins of eucalyptus wood, which is generally used for this purpose in California. Twelve of these pins were of a size much used in California on lines where the voltage is not above 30,000. Each of the twelve pins was $6\frac{7}{8}$ inches long in the stem, $4\frac{5}{8}$ inches long in the shank, $1\frac{1}{2}$ inches in diameter at the shank, 2 inches in diameter at the square shoulder where the shank joins the stem, and $1\frac{3}{8}$ inches in diameter at the top of the thread. The pins were tested by mounting each of them in a cross-arm, securing the cross-arm in a testing machine so that the pin was horizontal, placing an insulator on the pin, and exerting the strain on a cable wrapped around the side groove of the insulator. This cable varied a little from right angles to the axis of each pin, but the component of the strain at right angles to this axis was calculated and the breaking load here mentioned is that component. Nearly all of these twelve pins broke square off at the cross-arm.

For a single pin, the lowest breaking strain was 705 pounds, the largest 1,360 pounds, and the average for the twelve pins was 1,085 pounds. Unfortunately, the exact distance of the cable from the cross-arm is not stated, but as the cable was wound about the side groove of the insulator it was probably either in line with or a little below the top of the pin. It seems probable also that the diameter of these pins at the shoulder—that is, two inches—may have increased the breaking strain somewhat by giving the shoulder a good bearing on the cross-arm. The ten other pins were of the size in use on the 60,000-volt line between Colgate power-house and Oakland, Cal. Each of these pins had a length of 5⅜ inches and a maximum diameter of 2⅛ inches in the shank, and a length of 10⅜ inches in the stem, with a diameter of 2½ inches at the shoulder. This shoulder was not square, but its surface formed an angle of forty-five degrees with the axis of the pin, and this bevel shoulder took up ¼ inch of the length just given for the stem of the pin. At 2½ inches from its threaded end the stem of the pin had a diameter of 1 15/16 inches, and the diameter slopes to 1⅜ inches at two inches from the end. The two inches of length at the top of the stem has the uniform diameter of 1⅜ inches, and is threaded with four threads per inch for the insulator. Each of these ten pins was tested, as already described, until it broke, but the break in this case started as a split at the lower end of the threaded portion and ran down the stem to the shoulder in a line nearly parallel with the axis of the pin. The pull on the cable at right angles to the axis of each pin had a maximum value of 1,475 pounds in one case, and a corresponding value of 3,190 pounds in another, while the average breaking strain for the ten pins was 2,310 pounds. Unfortunately, the report of this test above named does not distinctly state just how far the testing cable was attached above the shank of each of these large pins; but it seems probable that the same insulator was used with the larger as with the smaller pins, and if this was so the testing cable was attached near the end of each pin, as this cable was wound about the side groove of the insulator used on the smaller pins. With the types of insulator in actual use on the Colgate and Oakland line the wire is carried at the top groove and its centre is about two and a half inches above the top of the pin. It is therefore probable that these pins would not withstand as great strains on the lines as they did in these tests. The bevel shoulder on each of these larger pins no doubt increases its ability to resist a bending strain, because the bevel surface fits tightly down into a counterbore in the cross-arm. Where the pin has a shoulder at right angles with the axis, as is more usually the case, and the top of the cross-arm is a little rounding, the square shoulder does not have a firm seat and is of slight importance as far as the strength of the pin to resist a bending strain is concerned. Evidently the weakest point in the ten larger pins of this test was at the lower end of the threaded portion, since in each case the break was in the form of a long split starting where the thread ended. There seems to be no sufficient reason for the reduction of the diameter of a pin intended for a heavy line wire to a diameter as small as one inch at the threaded end, or for limiting the length of the threaded portion to 2.5 inches, as proposed in the specifications for standard pins. It is certain that the cost of the pin would be no more if its diameter at the threaded end were 1¼ or 1⅜ inches with a uniform taper from the end of the pin down to the shoulder and with the thread cut down the stem for three or four inches. Furthermore, any increase in the cost of insulators for these larger threaded ends of pins would no doubt be a small matter. Some excess of strength in the stem of a pin over that of its shank is to be desired, for the stem is more exposed to the weather and to charring by leakage currents over the surface of the insulator. On high-voltage lines, this charring is usually worse at that part of each pin just below its thread, and the commonest breaks of pins on these lines leave the insulators with the threaded portions of their pins hanging on the wire, while the remainder of each pin remains on the cross-arm. From the tests just noted it is evidently poor design to give the threaded portion of a pin a short length of uniform diameter, and then to increase the diameter at once by a shoulder, as was done with the pins on the Colgate and Oakland line. This design evidently leads to failure of pins by splitting from the lower end of the threads. The better design is the more common one which gives the stem of the pin a uniform taper from the shoulder to the top. Where the line wire is secured to the top of its insulator, anywhere from one to three inches above the top of the pin, there is a strong tendency for the insulator to tip on its pin, and this tendency is more effectively met the longer the joint between the pin and insulator.

CHAPTER XXIII.
STEEL TOWERS.

Steel towers are rapidly coming into use for the support of electric transmission lines that deliver large units of energy at high voltages to long distances from water-powers.

One case of this sort is the seventy-five-mile transmission of 24,000 horse-power at 60,000 volts from Niagara Falls to Toronto. Another example may be seen in the seventy-five-mile line of steel towers which carries transmission circuits of 60,000 volts to Winnipeg. Guanajuato, Mexico, which is said to have produced more silver than any other city in the world, receives some 3,300 electric horse-power over a 60,000-volt transmission line one hundred miles long on steel towers. Between Niagara Falls and Lockport the electric circuits now being erected are supported on steel towers. On a transmission line eighty miles long in northern New York, for which plans are now being made, steel towers are to support electric conductors that carry current at 60,000 volts.

For the elevations above ground at which it is common to support the conductors of transmission lines—that is, from twenty-five to fifty feet—a steel tower will cost from five to twenty times as much as a wooden pole in various parts of the United States and Canada. It follows at once from this fact that there must be cogent reasons, apart from the matter of first cost, if the general substitution of steel towers for wooden poles on transmission lines is to be justified on economic grounds. During fifteen years the electric transmission of energy from distant water-powers to important centres of population has grown from the most humble beginnings to the delivery of hundreds of thousands of horse-power in the service of millions of people, and the lines for this work are supported, with very few exceptions, on wooden poles. Among the transmissions of large powers over long distances at very high voltages that have been in successful operation during at least several years with wooden pole lines are the following: the 60,000-volt circuit that transmits some 13,000 horse-power from Electra station across the State of California to San Francisco, a distance of 147 miles, is supported by wooden poles. In the same State, the transmission line 142 miles long between Colgate power-house and Oakland, at 60,000 volts, and with a capacity of about 15,000 horse-power, hangs on wooden poles, save at the span nearly a mile long over the Straits of Carquinez. Wood is used to carry the two 55,000-volt circuits that run sixty-five miles from the 10,000-horse-power station at Cañon Ferry on the Missouri River to Butte. Between Shawinigan Falls and Montreal, a distance of eighty-three miles, the conductors that operate at about 50,000 volts are carried on wooden poles.

Electrical supply in Buffalo to the amount of 30,000 horse-power depends entirely on circuits from Niagara Falls that operate at 22,000 volts and are supported on lines of wooden poles.

In the operation of these and many other high-voltage transmissions during various parts of the past decade some difficulties have been met with, but they have not been so serious as to prevent satisfactory service. Nevertheless, it is now being urged that certain impediments that are met in the operation of transmission systems would be much reduced by the substitution of steel towers for wooden poles, and it is even suggested that perhaps the first cost, and probably the last cost, of a transmission line would be less with steel than with wood for supports. The argument for steel in the matter of costs is that while a tower requires a larger investment than a pole, yet the smaller number of towers as compared with that of poles may reduce the entire outlay for the former to about that for the latter. More than this, it is said that the lower depreciation and maintenance charges on steel supports will make their final cost no greater than that of wooden poles.

In the present state of the market, steel towers can be had at from three to three and one-half cents per pound, and the cost of a steel tower or pole will vary nearly as its weight. During the first half of 1904 the quotations on tubular steel poles to the Southside Suburban Railway Company, of Chicago, were between the limits just stated. That company ordered some poles built up of steel sections about that time at a trifle less than three cents per pound. Each of these poles was thirty feet long and weighed 616 pounds, so that its cost was about eighteen dollars (xxi, A. I. E. E., 754). For a forty-five-foot steel pole to carry a pair of 11,000-volt, three-phase circuits along the New York Central electric road the estimated cost was eighty dollars in the year last named (xxi, A. I. E. E., 753). On the 100-mile line to Guanajuato, Mexico, above mentioned, the steel towers were built up of 3" × 3" × 3⁄16" angles for legs, and were stayed with smaller angle sections and rods. Each of these towers has four legs that come together near the top, is forty feet high, weighs about 1,500 pounds, and carries a single circuit composed of three No. 1 B. & S. gauge hard-drawn copper cables. The weight of each of these cables is 1,340 pounds per mile, and the forty-foot towers are spaced 440 feet apart, or twelve per mile, over nearly the entire length of line. At three cents per pound, the lowest figure at which these towers could probably be secured for use in the United States, the approximate cost of each would be forty-five dollars. Between Niagara Falls and Lockport each of the steel towers that is to carry a single three-phase transmission circuit has three legs built up of tubing that tapers from two and one-half inches to smaller sizes and is braced at frequent intervals. The height of these towers is forty-nine feet, and the weight of each is 2,800 pounds. At three cents per pound the cost of each tower amounts to eighty-four dollars. For a long transmission line in northern New York bids were recently had on towers forty-five feet high to carry six wires, and the resulting prices were $100 to $125 each for a tower weighing about 3,000 pounds. On the line between Niagara Falls and Toronto the standard tower holds the lowest cables 40 feet above ground at the insulators, has a weight of 2,360 pounds, and would cost $70.80 at 3 cents per pound.

In January, 1902, four steel towers were purchased to support transmission circuits for two spans of 132 feet each over the Chambly Canal, near Chambly Canton, Quebec. Each pair of these towers was required to support eleven No. 2-0 B. & S. gauge bare copper wires with the span of 132 feet between them. The vertical height of each of these four towers is 144 feet above the foundation, and they were designed for a maximum stress in any member of not more than one-fourth of its ultimate strength, with wires coated to a diameter of one inch with ice and under wind pressure. For these four steel towers erected on foundations supplied by the purchasers the price was $4,670, and the contract called for a weight in the four towers of not less than 121,000 pounds. On the basis of this weight the cost of the towers erected on foundations was 3.86 cents per pound.

With these examples of the cost of steel towers a fair idea may be gotten of the relative cost of wooden poles. For poles of cedar or other desirable wood thirty-five feet long and with eight-inch tops fitted with either one or two cross-arms an estimated cost of five dollars each is ample to cover delivery at railway points over a great part of the United States and Canada. This size of pole has been much used on the long, high-voltage transmission systems that involve large power units and use heavy conductors. Examples of lines where such poles are used may be seen between Niagara Falls and Buffalo, between Colgate power-house and Oakland, and between Cañon Ferry and Butte. Of course some longer poles were used in special locations, like the crossing of steam railways, but it is also true that on the lines supported by steel towers such locations make exceptionally high towers necessary. The thirty-five-foot poles will hold the electric lines about as high above the ground level as the forty-nine-foot towers on the Niagara Falls and Toronto transmission, because the former will be set so much closer together. On the line just named the regular minimum distance of the electric cables above the ground level at the centres of spans is twenty-five feet. The standard towers on this line carry the lower electric cables forty feet above the ground at the insulators, and it was thought desirable to allow a sag of fifteen feet at the centres of the regular spans of four hundred feet each. On these towers the conductors that form each three-phase circuit are six feet apart, and lines drawn between the three cables form the sides of an equilateral triangle. With a pin fourteen and three-fourths inches long like that used on these steel towers, and one conductor at the top of a thirty-five-foot pole, where the other two are supported by a cross-arm five feet three inches below, giving six feet between cables, the lower cables are held by their insulators twenty-six feet above the ground, when the poles are set five feet deep. Between thirty-five-foot poles one hundred feet is a very moderate span, and one that is exceeded in a number of instances. Thus on the 142-mile line from Colgate power-house to Oakland the thirty-five-foot poles are 132 feet apart, and one line of these poles carries three conductors of 133,000-circular-mil copper, while the other pole line has three aluminum cables of 168,000 circular mils. On the later transmission line from Niagara Falls to Buffalo, which was designed for three-phase circuits of 500,000-circular-mil cable, the regular distance between the thirty-five-foot poles is 140 feet.

A maximum sag of twenty-four inches between poles 100 feet apart under the conditions named above brings the lowest points of the wire twenty-four feet above the ground. The steel towers on the line to Guanajuato being only forty feet in length, and spaced 440 feet apart, it seems that the distance of conductors from the ground at the centres of spans is probably no greater than that just named. Particular attention is called to this point because it has been suggested that the use of steel towers would carry cables so high that wires and sticks could not be thrown onto them. It thus appears that thirty-five-foot wooden poles set one hundred feet apart will allow as much distance between conductors, and still keep their lowest points as far above the ground, as will forty- to forty-nine-foot towers placed four hundred feet or more apart. The two lines that have their conductors further apart perhaps than any others in the world are the one from Cañon Ferry to Butte, on thirty-five-foot wooden poles, and the one to Guanajuato, on steel towers. In each of these cases the cables are seventy-eight inches apart at the corners of an equilateral triangle. With steel towers four hundred feet or wooden poles one hundred feet apart, four of the latter must be used to one of the former. At $5 per pole this requires an investment of $20 in poles as compared with at least $45 for a tower like those on the Guanajuato line, $84 for a tower like those on the line from Niagara Falls to Lockport, or $70 for one of the towers on the Niagara and Toronto line. Each of the towers on the line to Toronto carries two three-phase circuits, and the least distance between cables is six feet. To reach the same result as to the distance between conductors with the two circuits on poles, it would be desirable to have two pole lines, so that $40 would represent the investment in the poles to displace one tower for two circuits. The older pole line between Niagara Falls and Buffalo carries two three-phase circuits on two cross-arms, and the 350,000-circular-mil copper cables of each circuit are at the angles of an equilateral triangle whose sides are each three feet long. In this case, however, the electric pressure is only 22,000 volts.

The costs above named for poles and towers include nothing for erection. Each tower has at least three legs and more commonly four, and owing to the heights of towers and to the long spans they support it is the usual practice to give each leg a footing of cement concrete. It thus seems that the number of holes to be dug for a line of towers is nearly or quite as great as that for a line of poles, and considering the concrete footings the cost of erecting the towers is probably greater than that for the poles. With wooden poles about four times as many pins and insulators are required as with steel towers, or say twelve pins and insulators on poles instead of three on a tower. For circuits of 50,000 to 60,000 volts the approximate cost of each insulator with a steel pin may be taken at $1.50, so that the saving per tower reaches not more than $13.50 in this respect. In the labor of erecting circuits there may be a small advantage in favor of the towers, but the weight of the long spans probably offsets to a large extent any grain of time due to fewer points of support.

An approximate conclusion from the above facts seems to be that a line of steel towers will probably cost from 1.5 to twice as much as a line or lines of wooden poles to support the same number of conductors the same distance apart, even when the saving of pins and insulators is credited to the towers. This conclusion applies to construction over a large part of the United States and Canada. It is known that wooden poles of good quality retain enough strength to make them reliable as supports during ten or fifteen years, and it is doubtful whether steel towers will show enough longer life to more than offset their greater first cost. It may be noted here that any saving in the cost of insulators or other advantage that there may be in spans four hundred feet or more long can be as readily secured with wooden as with steel supports. With these long spans the requirements are greater height and strength in the line supports, and these can readily be obtained in structures each of which is formed of three or four poles with cross-braces. Such wooden structures have long been in use at certain points on transmission lines where special long spans were necessary or where there were large angular changes of direction. In those special cases where structures 75 to 150 or more feet in height are necessary to carry a span across a waterway, as at the Chambly Canal above mentioned, steel is generally more desirable than wood because poles of such lengths are not readily obtainable. Neither present proposals nor practice, however, contemplates the use of steel towers having a length of more than forty to fifty feet on regular spans.

Much the strongest argument in favor of steel towers for transmission lines is that these towers give a greater reliability of operation than do wooden poles. It is said that towers will act as lightning-rods and thus protect line conductors and station apparatus. As to static and inductive influences from lightning, it is evident that steel towers can give no protection. If each tower has an especial ground connection it will probably protect the line to some extent against direct lightning strokes, but there is no reason to think that this protection will be any greater than that given by well-grounded guard wires, or even by a wire run from a ground plate to the top of each pole or wooden tower. If a direct lightning stroke passes from the line conductors to a wooden support it frequently breaks the insulator on that support, and the pole is often shattered or burned. Such a result does not necessarily interrupt the transmission service, however, as the near-by poles can usually carry the additional strain of the line until a new pole can be set. Quite a different result might be reached if lightning or some other cause broke an insulator on a steel tower, and thus allowed one of the electric cables to come into contact with the metal structure, as the conductor would then probably be burned in two. To repair a heavy cable thus severed where the spans were as much as 400 feet long would certainly require some little time. Where a conductor in circuits operating at 20,000 to 35,000 volts has in many cases dropped onto a wooden cross-arm, it has often remained there without damage until discovered by the line inspector, but no such result could be expected with steel towers and cross-arms (xxi, A. I. E. E., 760). Where steel towers are employed it would seem to be safer to use wooden cross-arms, for the reasons just stated. This is, in fact, the practice on the steel towers before named that support 25,000-volt circuits over the Chambly Canal, and also on the steel towers that carry the 60,000-volt circuits from Colgate power-house over the mile-wide Straits of Carquinez.

On the 40,000-volt transmission line between Gromo and Nembro, Italy, where timber is scarce and steel is cheap, both the poles and cross-arms are of wood. It is thought that the comparatively small number of insulators used where a line is supported at points about four hundred feet apart should contribute to reliability in operation, but insulators now give no more trouble than other parts of the line, and the leakage of energy over their surfaces is very small in amount, as was shown in the Telluride tests. Whatever benefits are to be had from long spans are as available with wooden as with steel supports, and at less cost.

One advantage of steel towers over wooden poles or structures is that the former will not burn and are probably not subject to destruction by lightning. Where a long line passes over a territory where there is much brush, timber or long grass, the fact that steel towers will not burn may make their choice desirable. In tropical countries where insects rapidly destroy wooden poles the use of steel towers may be highly desirable even at much greater cost, and such a case was perhaps presented on the line to Guanajuato, Mexico.

Mechanical failures of wooden insulator pins have been far more common than those of poles, both as a direct result of the line strains and because such pins are often charred and weakened by the leakage of energy from the conductors. For these reasons the general use of iron or steel pins for the insulators of long lines operating at high voltages seems desirable. Such pins are now used to support the insulators on a number of lines with wooden poles and cross-arms, among which may be mentioned the forty-mile, 30,000-volt transmission between Spier Falls and Albany and the forty-five-mile 28,000-volt line from Bear River to Ogden, Utah. Iron or steel pins add very little to the cost of a line, and materially increase its reliability. One of the cheapest and best forms of steel pins is that swaged from a steel pipe and having a straight shank and tapering stem with no shoulder. A pin of this sort for the 400-foot spans of 190,000-circular-mil copper cable on the line from Niagara Falls to Toronto measures three and one-quarter inches long in the shank, eleven and one-half inches in the taper, and has diameters of two and three-eighths inches at the larger and one and one-eighth inches at the smaller end. On spans under 150 feet between wooden poles pins of this type but with a much smaller diameter could be used to advantage.

On long transmission lines where the amount of power involved is very large the additional reliability to be had with steel towers is probably great enough to justify their use. For the great majority of power transmissions, however, it seems probable that wooden poles or structures will long continue to be much the cheaper and more practicable form of support.

The line of steel towers on a private right of way seventy-five miles long, carrying two circuits for the transmission of 24,000 horse-power at 60,000 volts from Niagara Falls to Toronto, is one of the most prominent examples of this type of construction.

Eventually there will be two rows of steel towers along the entire length of the line.

On the straight portions of the line the steel towers are regularly erected 400 feet apart, but on curves the distances are less between towers, so that their total number is about 1,400 for each line. Standard curving along the line requires towers placed 50 feet apart, and a change in the direction of not more than ten degrees at each tower, except at the beginning and end of the curve, where the change in direction is three degrees. When the change in the direction of the line is not more than six degrees, the corresponding spans allowed with each change are as follows:

Degrees change.		Feet of span.
	½	300
1		286
1	½	273
2		259
2	½	246
3		232
3	½	219
4		205
4	½	192
5		178
5	½	165
6		151

At some points along the line conditions require a span between towers of more than 400 feet, the regular distance for straight work. One example of this sort occurs at Twelve-Mile Creek, where the stream has cut a wide, deep gorge in the Erie plateau. At this point the lines make a span of 625 feet between towers.

FIG. 94.—Transposition Tower (Second Tower).

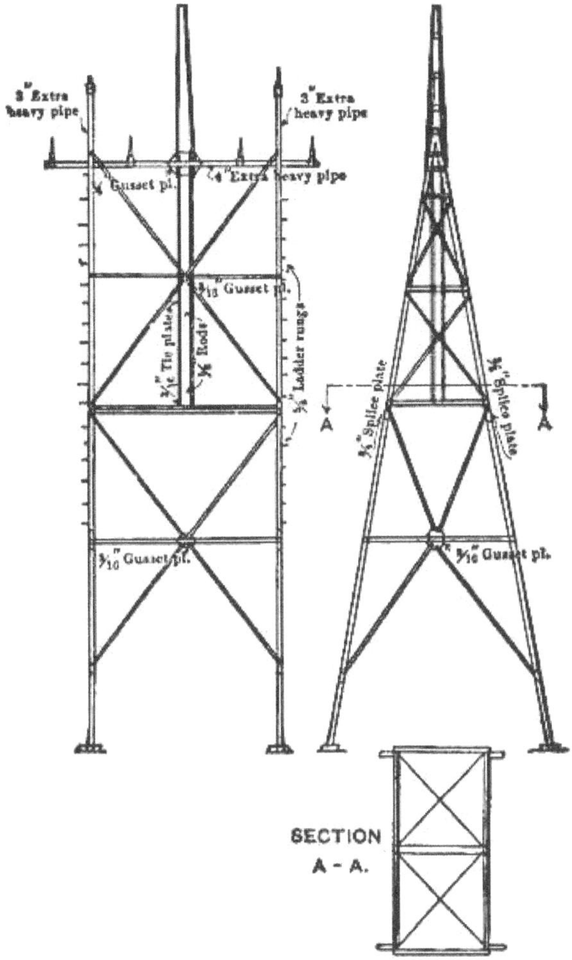

FIG. 95.—Elevations and Plan of Tower.

The regular steel tower used in this transmission measures 46 feet in vertical height from its foot to the tops of the lower insulators, and 51 feet 3 inches to the tops of the higher insulators. The lower six feet of this tower are embedded in the ground, so that the tops of the insulators measure about 40 feet and 45 feet 3 inches respectively above the earth. At the ground the tower measures 14 feet at right angles to the transmission line and 12 feet parallel with it. The width of each tower at the top is 12 feet at right angles to the line, and the two sides having this width come together at points about 40 feet above the ground. Between the two L bars thus brought nearly together, at each side of a tower a piece of extra heavy 3-inch steel pipe is bolted so as to stand in a vertical position. Each piece of this pipe is about

$3\frac{1}{2}$ feet long and carries a steel insulator pin at its upper end. The two pieces of pipe thus fixed on opposite sides of the top of a tower carry the two highest insulators. For the other four insulators of each tower, pins are fixed on a piece of standard 4-inch pipe that serves as a cross-arm, and is bolted in a horizontal position between the two nearly rectangular sides of each tower, at a point two feet below the bolts that hold the vertical 3-inch pipes, already named, in position. Save for the two short vertical and one horizontal pipe, and the pins they support, each tower is made up of L-shaped angle-bars bolted together. Each of the two nearly rectangular sides of a tower consists of two L bars at its two edges, three L bars for cross-braces at right angles to the edges, and four diagonal braces also formed of L bars. The lower halves of the L bars at the edges of each side of a tower have sections of 3" × 3" × $\frac{1}{4}$", and the upper halves have sections of 3" × 3" × $\frac{3}{16}$". This last-named cross-brace and the other two cross-braces have a common section of 2" × $1\frac{1}{2}$" × $\frac{1}{8}$". For the lower set of diagonal braces the common section is $2\frac{1}{2}$" × 2" × $\frac{1}{8}$", and the upper set has a section of 2" × $1\frac{1}{2}$" × $\frac{1}{8}$" in each member. At the level of the lowest cross-braces the two rectangular sides of a tower are tied together by one member of 2" × $1\frac{1}{2}$" × $\frac{1}{8}$" of L section and at right angles to the sides, and by two diagonal braces of $\frac{5}{8}$" round rod between the corners of the tower. On each of its two triangular sides a tower has four horizontal braces and three sets of diagonal braces. The two upper horizontal braces are of 2" × $1\frac{1}{2}$" × $\frac{1}{8}$" L section, and the lowest is the same, but the remaining horizontal brace has a section of $2\frac{1}{2}$" × 2" × $\frac{1}{8}$". Bars of 2" × $1\frac{1}{2}$" × $\frac{1}{8}$" L section are used for the two upper sets of diagonal braces, and bars of $2\frac{1}{2}$" × 2" × $\frac{1}{8}$" for the lower set. In addition to the cross-braces named, each triangular side of a tower near the top of the corner bars has two short cross-pieces with the common L section of $3\frac{1}{2}$" × $3\frac{1}{2}$" × $\frac{5}{8}$", one just above and the other just below the cross-arm of 4-inch pipe to hold it in place. At the bottom of each of the four corner bars of a tower a foot is formed by riveting a piece of 3" × $\frac{1}{4}$" L section and 15 inches long at right angles to the corner bar. On one corner bar of each tower there are two rows of steel studs for steps, one row being located in each flange of the L section. On the same flange these steps are two feet apart, but taking both flanges they are only one foot apart. Every part of each steel tower is heavily galvanized.

FIGS. 96, 97, 98.—Raising Towers on Niagara Transmission Line.

FIG. 99.—One of the Towers in Position.

The labor of erecting these steel towers was reduced to a low figure by the method employed, as shown in the accompanying illustration. Each tower was brought to the place where it was to stand with its parts unassembled. For erecting the towers a four-wheel wagon with a timber body about thirty feet long was used. When it was desired to raise a tower, two of the wheels, with their axle, were detached from the timber body of the wagon, and this body was then stood on end to serve as a sort of derrick. This derrick was guyed at its top on the side away from the tower, and a set of blocks and tackle was then connected to the top of the derrick and to the tower at a point about one-fourth of the distance from its top. A rope from this set of blocks ran through a single block fixed to the base of the derrick and then to a team of horses. On driving these horses away from the derrick the steel tower was gradually raised on the two legs of one of its rectangular sides until it came to a vertical position. The next operation was to bring the legs of the tower into contact with the extension pieces that were fixed in the earth and then bolt them together.

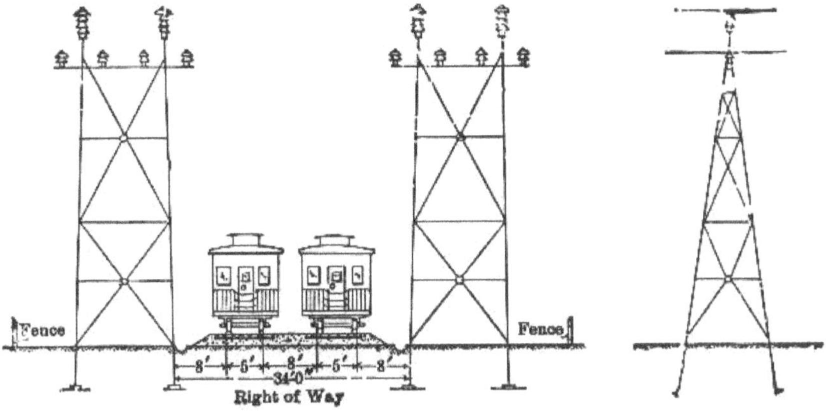

FIG. 100.—Steel Tower for Transmission Line.

The tops of the three pins that carry the insulators for each three-phase circuit are at the corners of an equilateral triangle (Fig. 100), each of whose sides measures six feet. The six steel insulator pins used on each tower are exactly alike, and each is swaged from extra heavy pipe. Each finished pin is $2\frac{3}{8}$ inches in diameter for a length of $3\frac{1}{4}$ inches, and then tapers uniformly to a diameter of $1\frac{1}{8}$ inch at the top through a length of $11\frac{1}{2}$ inches. This gives the pin a total length of $14\frac{3}{4}$ inches. In the larger part there are two $\frac{9}{16}$-inch holes from side to side, and within two inches of the top there are three circular grooves each $\frac{3}{16}$ inch wide and $\frac{1}{16}$ inch deep. Forged steel sockets of two types are employed to attach the steel pins with the pipes. Each socket is made in halves, and these halves are secured to both the pipe and the pin by through bolts. Like all other parts of the towers, these steel pins and sockets are heavily galvanized. On each of the four corner bars of a tower the lower six feet of its length is secured to the upper part by bolts or rivets. This lower six feet of each corner bar is embedded in the earth, and the construction just named makes it easy to replace the bars in the earth when corrosion makes it necessary.

Footings for each tower are provided by digging four nearly square holes with their sides at approximately 45 degrees with the direction of the transmission line, and the shortest side of each hole at least two feet long.

Centres of these holes are 14 feet 3 inches apart in a direction at right angles to the line, and 13 feet 9 inches apart parallel with the line. In hard-pan each one of the holes was filled to within 2 feet 6 inches of the top with stones, after the leg of the tower was in position, and then the remainder of the hole was filled with cement grouting mixed four to one.

At the bottom of each hole in marsh land a wooden footing 3 feet × 6 inches × 24 inches was laid flat beneath the leg of the tower, and then the hole was filled to within 2½ feet of the surface with the excavated material. Next above this filling comes a galvanized iron gutter-pipe, four inches in diameter, and filled with cement about the leg of the tower for a length of two feet. Outside of this pipe the hole is made rounding full of cement grouting.

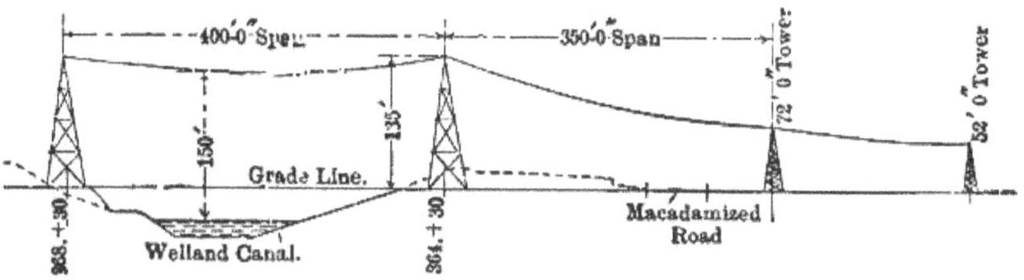

FIG. 101.—Transmission Line at Welland Canal.

At some points along the transmission line exceptionally high towers are necessary, a notable instance being found at the crossing over the Welland Canal, where the lowest part of each span must not be less than 150 feet above the water. For this crossing two towers 135 feet high above ground are used, as seen in Fig. 101. Each of these towers is designed to carry all four of the three-phase power circuits that are eventually to be erected between Niagara Falls and Toronto. For this purpose there was used a special design of tower with a width of about 48 feet at right angles to the direction of the line below the top truss, and a width of about 68.5 feet at that truss where the two lower conductors of each circuit are attached.

With all spans longer than 400 feet, a tower of heavier construction than the standard type is used, and this tower provides three insulators for the support of each conductor. A tower of this type that supports the lowest conductors about 40 feet above the ground level has its corner bars made up of 4" × 4" × 3/8" and 4" × 4" × 5/16" L sections, has three cross-arms of extra heavy 4-inch pipe, and a 6-inch vertical standard pipe to support each group of three insulators for the highest conductor of each circuit. Each of the lower conductors of a circuit on this tower is supported by an insulator on each of the three parallel cross-arms. On some of these towers, for long spans, the two outside insulators for the support of each conductor are set a little lower than the insulator between them.

FIG. 102.—Heavy Tower at Credit River.

FIG. 103.—Angle Tower Near Bronte.

Angle towers, used where the line makes a large change in direction at a single point, have three legs on each rectangular side, a width of 20 feet on each of these sides for some distance above the ground, and a width of 27 feet 2 inches at the top. In these towers the two legs on the triangular side that is in compression are each made up of four 3" × 3" × ¼" L sections joined by 1½" × ¼" lattices and rivets. Towers of this sort are used near the Toronto terminal-station, where the line changes 35 degrees at a single point, and near the crossing of Twelve-Mile Creek, where the angular change of the line on a tower is 45 degrees. Close to each terminal-station and division-house the transmission line is supported by terminal towers. These towers differ from the others in that each carries insulators for only three conductors, and these insulators are all at the same level. Each terminal tower has nine insulators, arranged in three parallel rows of three each for the conductors of a single circuit, and each conductor thus has its strain distributed between three pins. All three wires of a circuit are held 40 feet above the ground by a terminal tower, and pass to their entries in the wall of a station at the same level. As these terminal towers must resist the end strain of the line, they are made extra heavy, the four legs each being made up of 4" × 4" × 5⁄16" and 4" × 4" × ⅜" L sections. For the three cross-arms on one of these towers three pieces of 4-inch pipe, each 15 feet 9 inches long, are secured at its top with their parallel centre lines 30 inches apart in the same plane. Each of these pipes carries three insulator pins with their centres 7 feet 4½ inches apart. On the bottom of each leg of a terminal tower there is a foot, formed by riveting on bent plates, that measure 15 and 18 inches, respectively, on the two longer sides. Each foot of this tower is set in a block of concrete 5 feet square that extends from 3.5 feet to 7.5 feet below the ground level.

Insulators for the transmission line, which are illustrated in Fig. 104, are of brown, glazed porcelain, made in three parts, and cemented together. The three parts consist of three petticoats or thimbles, each of which slips over or into one of the others, so that there are three outside surfaces and three interior or protected surfaces between the top of an insulator and its pin.

From top to bottom the height of each insulator is 14 inches, and this is also the diameter of the highest and largest petticoat. The next or middle petticoat has a maximum diameter of 10 inches and the lowest petticoat one of 8 inches. Cement holds the lowest petticoat of the insulator on one of the steel pins previously described, and in this position the edge of the lowest petticoat is about 2½ inches from the steel support. At the top of each insulator the transmission conductor is secured, and the shortest distance from this conductor to any of the steel parts through the air is about 17 inches.

From the step-up transformer house at Niagara Falls to the terminal-station at Toronto, a distance of seventy-five miles, each three-phase, 60,000-volt, 25-cycle circuit on the steel towers is made up of three hard-drawn copper cables with a cross section of 190,000 circular mils each, and is designed to deliver 12,000 electric horse-power with a loss of ten per cent, on a basis of 100 per cent power factor. Six equal strands of copper make up each cable, and this wire has been specially drawn with an elastic limit of more than 35,000 pounds and a tensile strength of over 55,000 pounds per square inch. This cable is made in uniform lengths of 3,000 feet, and these lengths are joined by twisting their ends together in copper sleeves, and no solder is used. No insulation is used on these cables.

FIG. 104.—Insulators.

Instead of a tie-wire, a novel clamp is employed to secure the copper cable on each insulator. This complete clamp is made up of two separate clamps that grasp the cable at opposite sides of each insulator and of two half-circles of hard-drawn copper wire of 0.187 inch diameter. Each half-circle of this wire joins one-half of each of the opposite clamps, and fits about the neck of the insulator just below its head. Two bronze castings, one of which has a bolt extension that passes through the other, and a nut, make up each separate clamp. When the combined clamp is to be applied, the sides are separated by removing the nut that holds them together, the half-circles are brought around the neck of the insulator, and each of the side clamps is then tightened on to the cable by turning the nut that draws its halves together. This complete clamp can be applied as quickly as a tie-wire, is very strong, and does not cut into the cable.

Each of the regular steel towers is designed to withstand safely a side strain of 10,000 pounds at the insulators, or an average of 1,666 pounds per cable. With the 190,000-mil cable coated to a depth of ½ inch with ice and exposed to a wind blowing 100 miles per hour, the estimated strains on each steel pin for different spans and angular changes in the direction of the line are given in the accompanying table:

POUNDS STRAIN ON PINS, ½-INCH SLEET, 100 MILES WIND.

Span, feet.	Degrees and Minutes.												
	0	0.30	1	1.30	2	2.30	3	3.30	4	4.30	5	5.30	6
0	0	35	69	104	138	173	207	242	276	311	345	380	414
100	256	291	325	360	394	429	463	498	532	567	601	636	670
200	512	547	581	616	650	685	719	754	788	823	857	892	926
300	768	803	837	872	906	941	975	1,010	1,044	1,079	1,113	1,148	1,182
400	1,024	1,059	1,093	1,128	1,162	1,197	1,231	1,266	1,300	1,335	1,369	1,404	1,438
500	1,280	1,315	1,349	1,384	1,418	1,453	1,487	1,522	1,556	1,591	1,625	1,660	1,694
600	1,536	1,571	1,605	1,640	1,674	1,709	1,743	1,778	1,812	1,847	1,881	1,916	1,950
700	1,792	1,827	1,861	1,896	1,930	1,965	1,999	2,034	2,068	2,103	2,137	2,172	2,206
800	2,048	2,083	2,117	2,152	2,186	2,221	2,255	2,290	2,324	2,359	2,393	2,428	2,462
900	2,304	2,339	2,373	2,408	2,442	2,477	2,511	2,546	2,580	2,615	2,649	2,684	2,718
1,000	2,560	2,595	2,629	2,664	2,698	2,733	2,767	2,802	2,836	2,871	2,905	2,940	2,974

The copper cables were so strung as to have a minimum distance from the ground of 25 feet at the lowest points of the spans. In order to do this the standard steel towers that hold the lower cables 40 feet above the ground level at the insulators are spaced at varying distances apart, according to the nature of the ground between them. At each tower the upper cable of each circuit is 5 feet 3 inches higher than the two lower cables, and this distance between the elevations of the upper and the lower cables is maintained whatever the amount of sag at the centre of each span. If there is a depression between two standard towers on a straight portion of the line, the sag in the centre of a span 400 feet long may be as much as 18 feet. Where a rise and fall in the ground between towers make it necessary to limit the sag to 14 feet in order to keep the lowest cables 25 feet above the highest point of earth, the length of span is limited to 350 feet. If the rise and fall of ground level between towers allow a sag of only 11 feet with the lowest cable 25 feet above the earth, the length of span with 40-foot towers is reduced to 300 feet; and if for a like reason the sag is limited to 8 feet, the span may only be 250 feet.

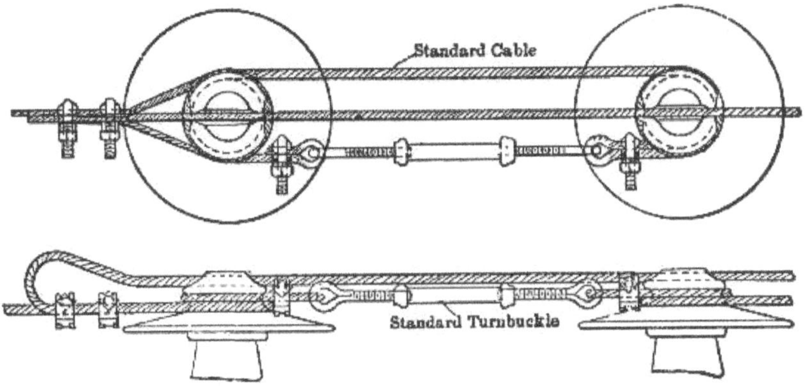

FIG. 105.—Take-up Arrangement on Terminal Tower.

At each terminal tower, where the cables are secured before they pass into a terminal-station, the three insulators for each cable are in a straight line with their centres, 30 inches apart. When a line cable reaches the first insulator of the three to which it is to be attached on one of these towers, it is passed around the neck of this insulator and then secured on itself by means of two clamps that are tightened with bolts and nuts. See Fig. 105. The cable thus secured turns up and back over the tops of the three insulators and goes to the terminal-station. Around the neck of the insulator to which the line cable has been secured in the way just outlined a short detached length of the regular copper cable with the parts of a turnbuckle at each end is passed, and this same piece of cable also passes around the neck of the next insulator in the series of three. By joining the ends of the turnbuckle and tightening it, a part of the strain of the line cable in question is transferred from the first to the second insulator of the series. In the same way a part of the strain of this same line cable is transferred from the second insulator of the series to the third, or one nearest to the terminal-station.

Made in United States
Cleveland, OH
05 November 2026

47667152R00185